Weyl Group Multiple Dirichlet Series: Type A Combinatorial Theory

Ben Brubaker, Daniel Bump, and Solomon Friedberg

PRINCETON UNIVERSITY PRESS

PRINCETON AND OXFORD

2011

Copyright © 2011 by Princeton University Press
Published by Princeton University Press, 41 William Street, Princeton, New Jersey 08540

In the United Kingdom: Princeton University Press, 6 Oxford Street, Woodstock, Oxfordshire OX20 1TW

press.princeton.edu

Library of Congress Cataloging-in-Publication Data

Brubaker, Ben, 1976–
Weyl group multiple Dirichlet / Ben Brubaker, Daniel Bump, and Solomon Friedberg.
    p. cm. – (Annals of mathematics studies ; no. 175)
Includes bibliographical references and index.
ISBN 978-0-691-15065-9 (hardcover : alk. paper) – ISBN 978-0-691-15066-6 (pbk.: alk. paper) 1. Dirichlet series. 2. Weyl groups. I. Bump, Daniel, 1952– II. Friedberg, Solomon, 1958– III. Title. IV. Series.

QA295.B876 2011
515'.243–dc22

2010037073

British Library Cataloging-in-Publication Data is available

This book has been composed in LaTeX

The publisher would like to acknowledge the authors of this volume for providing the camera-ready copy from which this book was printed.
Printed on acid-free paper. ∞
Printed in the United States of America
10 9 8 7 6 5 4 3 2 1

Contents

Preface

An *L-function*, as the term is generally understood, is a Dirichlet series in one complex variable s with an Euler product and (at least conjecturally) an analytic continuation to a meromorphic function on the whole complex plane and functional equation under the simple reflection $s \mapsto 1 - s$.

By contrast *Weyl group multiple Dirichlet series* are a new class of arithmetically interesting Dirichlet series that differ from L-functions in two ways. First, the coefficients of the series are only multiplicative up to a root of unity given in terms of power residue symbols. We refer to this property as *twisted multiplicativity*, given precisely in (1.3) in Chapter 1. This condition guarantees that the global series is determined by its values at prime-power supported coefficients, or *p-parts* for short. Second, they may be Dirichlet series in several complex variables $s_1, \cdots, s_r$ for any positive r. The series have (at least conjecturally) meromorphic continuation to all $\mathbb{C}^r$ with functional equations whose action on $\mathbb{C}^r$ is a finite reflection group.

The data needed to define a family of such series in r complex variables are a root system Φ of rank r with Weyl group W, a fixed integer $n \geqslant 1$, and a global ground field F containing the n-th roots of unity; in fact, it is much more convenient to assume that the ground field F contains the $2n$-th roots of unity so we make this assumption throughout, though it is not strictly necessary.

Weyl group multiple Dirichlet series appear naturally in the Fourier-Whittaker coefficients of Eisenstein series on the metaplectic covers of reductive groups. Their p-parts, the key objects of study, are metaplectic spherical Whittaker functions over a local field. The rank one case was studied by Kubota [53]. He showed that a one-variable Dirichlet series whose coefficients are each a single n-th order Gauss sum has analytic continuation and functional equations coming from the theory of Eisenstein series on an n-fold metaplectic cover of SL_2. For higher rank, it is not hard to show that the Whittaker coefficients of Eisenstein series are Dirichlet series in several complex variables. However, in the higher rank cases this interpretation as a Whittaker coefficient does not readily allow one to determine the multiple Dirichlet series coefficients or study their combinatorial and representation-theoretic properties.

In this book, we take an alternate approach to the study of Weyl group multiple Dirichlet series. In the special case where the root system Φ is of Cartan Type A, we present several explicit definitions of their p-parts as sums of number-theoretic quantities (products of n-th order Gauss sums) indexed by combinatorial data. It turns out that if one knows enough about the combinatorics of the coefficients, one may reduce the analytic continuation and functional equations to the rank one case

(Kubota's theory) by use of Bochner's tube domain theorem [7]. This approach was initially outlined in Bump, Friedberg, and Hoffstein [20] and detailed further in Brubaker, Bump, Chinta, Friedberg, and Hoffstein [10]. The combinatorics of the p-parts of Weyl group multiple Dirichlet series were thus studied independently of their realization as Whittaker coefficients of higher-rank metaplectic Eisenstein series, even though this connection was hypothesized from the very beginning.

Though Whittaker coefficients of higher rank groups are not, strictly speaking, used in the methods of this book, they are always in the background. The combinatorial definition for the p-part we will present in Chapter 1 was conjectured based on evidence from an SL_3 Whittaker computation made in Brubaker, Bump, Friedberg, and Hoffstein [17] and the form of multiple Dirichlet series appearing in the earlier work [12]. Once equipped with this definition, the authors have shown that it matches the Whittaker coefficient of a metaplectic Eisenstein series on SL_{r+1} in [13] by proving that both satisfy the same recursive relation. Moreover, McNamara [65] has shown that one of the two p-part definitions of Chapter 1 is equal to the p-adic metaplectic spherical Whittaker function.

In the nonmetaplectic case ($n = 1$), the spherical Whittaker functions are well-known. The Casselman-Shalika formula identifies these p-adic Whittaker functions for unramified principal series with characters of highest weight representations on the Langlands dual group (see Chapter 4 for more details). Based in part on the results of this book, we have come to view the p-parts as generalizations of characters of finite-dimensional irreducible representations of Lie groups in which the weights of the representation are modified by n-th order Gauss sums, where n is the degree of the metaplectic cover.

In Chapter 1, we make the preceding observation precise – giving explicit expressions for p-parts in Type A in terms of combinatorial bases for highest weight representations of the dual group $GL_{r+1}(\mathbb{C})$. The defining data for a multiple Dirichlet series encodes a dominant weight λ in the weight lattice of $GL_{r+1}(\mathbb{C})$ for each prime p. Given λ, we present two distinct definitions for the p-parts in terms of Gelfand-Tsetlin patterns that parametrize a basis for the representation with highest weight $\lambda + \rho$, where ρ is the usual Weyl vector.

The equivalence of these two descriptions is a deep fact that uses subtle combinatorial manipulations depending in an essential way on the properties of n-th order Gauss sums. This equivalence is the key step in demonstrating functional equations. This result is our main theorem and will occupy a large portion of the book.

While Gelfand-Tsetlin patterns provide a convenient language for the combinatorial proof of functional equations, the two definitions of the p-part as sums over patterns appear *ad hoc*. We remedy this in Chapters 2 and 3, where this pair of definitions is reformulated using alternate bases for the highest weight representation (the *string bases*) due to Berenstein and Zelevinsky [4], [5] and Littelmann [57]. Each string basis is given in terms of the Kashiwara crystal graph $\mathcal{B}_{\lambda+\rho}$ associated to highest weight $\lambda + \rho$ as follows. The basis depends on a decomposition of minimal length of the long element w_0 of the Weyl group into a product of simple reflections. Once this choice is made, there is a canonical path from each vertex of the crystal to the vertex corresponding to the lowest weight vector obtained by applying the root operators in the order appearing in the long word. The lengths of the resulting strings ("straight-line" segments of this path in the sense of Figures 2.1 and 2.2) are the basic data from which the p-parts are determined. In this context, the two Gelfand-Tsetlin descriptions from Chapter 1 correspond to a choice of one or the other of two particularly nice long words. There are many long words, but the two we need are in some sense as "far apart" as possible; for example, they are the first and last such decompositions in the lexicographical order.

In the case $n = 1$, the formula for the p-part as a generating function on the crystal (or equivalently on Gelfand-Tsetlin patterns) should match the Casselman-Shalika formula – that is, a character of a highest weight representation on the dual group. As a purely combinatorial identity, this statement is due to Tokuyama [70]. It may be viewed as a deformation of the Weyl character formula. In Chapter 5, we present Tokuyama's theorem in crystal language and explain how our definition of the p-part degenerates to this identity in the case $n=1$.

Chapters 6 through 17 prove the main theorem of the book. This asserts that the two Gelfand-Tsetlin descriptions for p-parts of multiple Dirichlet series defined in Chapter 1 agree. If this equivalence is known, then the analytic continuation and functional equations of the multiple Dirichlet series can be established by use of Bochner's theorem, as is described in detail in [15]. The argument is summarized in Chapter 6. If r is the rank, one needs to demonstrate functional equations corresponding to the r simple reflections that generate the Weyl group. Roughly speaking, one representation gives the *first* $r - 1$ simple reflections by induction from the rank one case, and the other gives the *last* $r - 1$ reflections.

Chapter 6 also outlines the proof, introducing many concepts and ideas, and several equivalent forms of the result, called Statements A through G. Each statement is an intrinsically combinatorial identity involving products of Gauss sums, but with each statement the nature of the problem changes. The first reduction (Statement B) changes the focus from Gelfand-Tsetlin patterns to "short" Gelfand-Tsetlin patterns, consisting of just three rows. This reduction, based on the Schützenberger involution, is explained in Chapters 7 and 18. A particular phenomenon called *resonance* is isolated, which seems to be at the heart of the difficulty in all the proofs. In Chapters 8 through 13, a reduction to the totally resonant case is accomplished. Now the equality is of two different sums of products of Gauss sums attached to the lattice points in two different polytopes. On the interiors of the polytopes, the terms bijectively match but on the boundaries a variety of perplexing phenomena occur. Moreover, the polytopes are irregular in shape. It is shown in Chapter 14 by means of an inclusion-exclusion process that these sums can be replaced by sums over lattice points in simplices; the terms that are summed are not the original products of Gauss sums but certain alternating sums of these. Although these terms appear more complicated than the original ones, they lead to an intricate but explicit rule for matching terms in each sum. This results in a final equivalent version of our main theorem, called Statement G, which is formulated in Chapter 15. This is proved in Chapters 16 through 17.

Once the proof of the equivalence of the two Gelfand-Tsetlin definitions is complete, we turn our attention to other matters. In Chapters 18 and 19, we reconsider Statement B from two different points of view. We will see that this deep but purely combinatorial statement has alternative formulations using other combinatorial models. The first reformulation of Statement B is in Chapter 18. Here we show that certain elements of the crystal, which we call "short vectors," are in bijection with short Gelfand-Tsetlin patterns. In this reinterpretation, the combinatorial data that we extract from a short Gelfand-Tsetlin pattern can alternatively be found by comparing data from two different paths from the short vector to the lowest weight vector in the crystal.

In Chapter 19 we then turn to the second alternative interpretation, using statistical mechanical methods to study invariance properties of the p-parts of Type A multiple Dirichlet series, or equivalently, of the spherical Whittaker function. The statistical mechanical system under consideration consists of a planar graph in the shape of a grid. A *state* of the system consists of an assignment of "spins" $\pm$ to the edges of the graph. Summing a statistic of the state – its Boltzmann weight – over the states of the system gives the *partition function*. This is a basic object of study in statistical mechanics. Certain systems are exactly solvable, meaning that the partition function may be computed explicitly. Baxter gave a method of obtaining such results using the Yang-Baxter equation to prove the commutativity of *row transfer matrices*, a phenomenon that he identified as the key to the solvability of the model.

Remarkably, one may formulate Statement B in terms of the commutativity of transfer matrices. Moreover, in the nonmetaplectic case ($n = 1$), one may follow Baxter in proving this fact using the Yang-Baxter equation. In this book, we will discuss the nonmetaplectic case completely, referring to Brubaker, Bump, Chinta, Friedberg, and Gunnells [8] for the metaplectic case. In that paper it is shown that, in addition to Statement B, a second property of the p-part, namely its transformation property under the action of the Weyl group on the Langlands parameters, may also be formulated in terms of the commutativity of transfer matrices.

The equivalence of the statistical mechanical and crystal descriptions of the p-parts follows from a bijection of the set of all admissible states of the model and a *subset* of the crystal $\mathcal{B}_{\lambda+\rho}$. This subset is precisely the set of elements on which the function describing the p-part is nonzero for some n. Going from the crystal description to the statistical mechanical one could *seem* to be merely a change of language, but in fact there is a subtle and important change of viewpoint involved. In switching to the statistical-mechanical picture, one loses one tool, namely the Schützenberger involution, which does not preserve the distinguished subset of the crystal. But one gains another tool: the Yang-Baxter equation. This is a local identity that, if true for a given choice of Boltzmann weights for the model, immediately implies the commutativity of transfer matrices (i.e., Statement B).

The final chapter of the book is devoted to Kashiwara's crystal $\mathcal{B}(\infty)$, which is a crystal basis of the enveloping algebra of the maximal unipotent subgroup. One can apply the definitions of the p-part to this crystal, and the philosophy here is that one may sometimes substitute an integral over the maximal unipotent subgroup by a summation over $\mathcal{B}(\infty)$.

We turn now to a discussion of the broader state of affairs in the theory of Weyl group multiple Dirichlet series, including matters that are largely beyond the scope of this book.

There is a second method for constructing the p-parts of multiple Dirichlet series due to Chinta and Gunnells [28, 29]. They give an action of the Weyl group on rational functions which, upon averaging over the Weyl group, produces a p-part with the correct invariance properties. Thus it may be viewed as a modification of the Weyl character formula. The form of the action was motivated in part by considerations from the function field case, and by worked-out examples such as the one in Chinta [24]. It is related to formulae originally described by Kazhdan and Patterson [49] in the context of metaplectic forms.

By averaging the action on a constant function, the method is known to produce metaplectic Whittaker functions by the work of Chinta and Offen [31], based on the method of Casselman and Shalika [23]. Thus, since the averaging method and the combinatorial basis methods of this book are both proved equal to the metaplectic spherical Whittaker function in Type A, they must agree. However, there is no combinatorial proof of this fact. (For partial progress in this direction, see Chinta, Friedberg, and Gunnells [26].)

The crystal graph description of multiple Dirichlet series suggests there should be an expression for the p-part corresponding to any choice of Weyl group and a reduced expression for the long element w_0. Our proof demonstrates equivalent definitions of the multiple Dirichlet series for several additional decompositions of the long word along the way, and the work of McNamara [65] gives an algorithm for explicitly describing the p-part to every long word.

However, this algorithm is difficult to execute outside of a few very special cases. At this writing, the crystal graph approach can be used to give a closed form answer for the p-part of Weyl group multiple Dirichlet series for at least one choice of long word in the following cases beyond Cartan Type A (all $n \geqslant 1$), as in the present document. For Cartan Types C (n odd), B (n even), and D (all n) there exist conjectural representations. These are rigorously proved in Type C for $n = 1$ and in Type B for $n = 2$. See Beineke, Brubaker, and Frechette [2, 3] for Type C, Brubaker, Bump, Chinta, and Gunnells [11] for Type B, and Chinta and Gunnells [30] for Type D. These representations for other classical groups may also be studied using statistical lattice models, as demonstrated by Ivanov in [44]. But a major open problem in the field remains to give a unified approach to defining p-parts corresponding to all root systems and all long words using statistics from the crystal graph.

Though our focus in this book is largely on the local components of Weyl group multiple Dirichlet series in the context of number fields, similar constructions work equally well over any global field containing the $2n$-th roots of unity. The functional equations established here imply that the multiple Dirichlet series over a global function field are rational functions in r variables. These rational functions have not been computed for higher rank, but several interesting phenomena have been observed in Cartan types A_2 and A_3 (cf. Hoffstein [41], Fisher and Friedberg [33, 34], and Chinta [25]).

One can seek to extend the theory of multiple Dirichlet series to the broader setting of Kac-Moody Lie algebras and their Weyl groups. A first case of multiple Dirichlet series having infinite group of functional equations – the affine Weyl group $D_4^{(1)}$ in Kac's classification – may be found in the work of Bucur and Diaconu [18]. Their result requires working over the rational function field and builds on work of Chinta and Gunnells [29]. If one could establish the analytic properties of such series in full generality, one would have a potent tool for studying moments of L-functions, as suggested in Bump, Friedberg, and Hoffstein [20].

If one tries to envision how the theory developed here might work for infinite Kac-Moody Weyl groups, one possibility is that Weyl group multiple Dirichlet series will be realized as Whittaker functions of Eisenstein series on loop groups, or "metaplectic" central extensions of loop groups as considered by Garland and Zhu [37]. Such an approach will obviously require deep foundational work. Alternatively, it may be fruitful also to pursue a combinatorial approach based on crystal graphs and making use of Bochner's convexity theorem. The methods of this book may serve as the basis for such an approach.

There are also other classes of multiple Dirichlet series. Some arise as residues of the series studied in this book. For example, Brubaker and Bump [8] indicated that one may take a multiple residue of the series on the n-fold cover of either A_n or A_{2n-2} and obtain a sum of twisted $GL(1)$ L-functions (with a proof for $n = 3$), a sum first studied as a multiple Dirichlet series by Friedberg, Hoffstein, and Lieman [36]. The full residue of the Type A multiple Dirichlet series for general n and r is also of interest; it is the Whittaker coefficient of a metaplectic theta function. Consideration of such functions for the triple cover of A_1 led to Patterson and Heath-Brown's disproof of the Kummer Conjecture [40]. Foundational work for the n-fold cover of A_r has been carried out by Kazhdan and Patterson [49], but the nature of these coefficients is still highly mysterious.

Additional classes of multiple Dirichlet series may be constructed from families of twisted automorphic L-functions (cf. Chinta, Friedberg, and Hoffstein [27] and Friedberg [35] and the papers referred to in those articles). Though we will not discuss such series in detail here, we mention that some of these may be analytically continued using Bochner's theorem. It is likely that many such series arise as the Whittaker coefficients on nonminimal parabolic metaplectic Eisenstein series or as the residues of such.

We would like to thank Gautam Chinta and Paul Gunnells for sharing their insights and papers in progress, and in particular for pointing out the relevance of Littelmann [57]. We thank Maki Nakasuji and Peter McNamara for many helpful conversations. In particular, Nakasuji helped with material in Chapter 5. Calculations made with Jeffrey Hoffstein were a source of inspiration for this work, and we thank him for years of shared ideas. We also thank Dorian Goldfeld, David Kazhdan, and S. J. Patterson for their interest and encouragement. Both SAGE and Mathematica were used in our investigations and SAGE (with its excellent support for crystals) was used in the preparation of the figures. This work was supported by grants from the NSF[1] and the NSA.[2]

---

[1]DMS-0652609, DMS-0652817, DMS-0652529, DMS-0702438, DMS-0844185, DMS-1001079, and DMS-1001326.
[2]H98230-07-1-0015 and H98230-10-1-0183.

Weyl Group Multiple Dirichlet Series

Chapter One

Type A Weyl Group Multiple Dirichlet Series

We begin by defining the basic shape of the class of Weyl group multiple Dirichlet series. To do so, we choose the following parameters.

- Φ, a reduced root system. Let r denote the rank of Φ.

- n, a positive integer,

- F, an algebraic number field containing the group μ_{2n} of $2n$-th roots of unity,

- S, a finite set of places of F containing all the archimedean places, all places ramified over $\mathbb{Q}$, and large enough so that the ring

$$\mathfrak{o}_S = \{x \in F \mid |x|_v \leqslant 1 \text{ for } v \notin S\}$$

  of S-integers is a principal ideal domain,

- $\boldsymbol{m} = (m_1, \cdots, m_r)$, an r-tuple of nonzero S-integers.

We may embed F and $\mathfrak{o}_S$ into $F_S = \prod_{v \in S} F_v$ along the diagonal. Let $(d, c)_{n,S}$ denote the S-Hilbert symbol, the product of local Hilbert symbols $(d, c)_{n,v} \in \mu_n$ at each place $v \in S$, defined for $c, d \in F_S^\times$. Let $\Psi : (F_S^\times)^r \longrightarrow \mathbb{C}$ be any function satisfying

$$\Psi(\varepsilon_1 c_1, \cdots, \varepsilon_r c_r) = \prod_{i=1}^{r} (\varepsilon_i, c_i)_{n,S} \prod_{1 \leq j < k \leq r} (\varepsilon_j, c_k)_{n,S}^{-1} \Psi(c_1, \cdots, c_r) \quad (1.1)$$

for any $\varepsilon_1, \cdots, \varepsilon_r \in \mathfrak{o}_S^\times (F_S^{\times,n})$ and $c_1, \cdots, c_r \in F_S^\times$. Here $(F_S^{\times,n})$ denotes the set of n-th powers in $F_S^\times$. It is proved in [12] that the set $\mathcal{M}$ of such functions is a finite-dimensional (nonzero) vector space.

To any such function Ψ and data chosen as above, Weyl group multiple Dirichlet series are functions of r complex variables $\boldsymbol{s} = (s_1, \cdots, s_r) \in \mathbb{C}^r$ of the form

$$Z_\Psi^{(n)}(\boldsymbol{s}; \boldsymbol{m}; \Phi) = Z_\Psi(\boldsymbol{s}; \boldsymbol{m}) = \sum_{\substack{c = (c_1, \cdots, c_r) \in (\mathfrak{o}_S / \mathfrak{o}_S^\times)^r \\ c_i \neq 0}} \frac{H(c; m) \Psi(c)}{\mathrm{Nc}_1^{2s_1} \cdots \mathrm{Nc}_r^{2s_r}}, \quad (1.2)$$

where Nc is the cardinality of $\mathfrak{o}_S / c \mathfrak{o}_S$, and it remains to define the coefficients $H(c; m)$ in the Dirichlet series. In particular, the function Ψ is not independent of the choice of representatives in $\mathfrak{o}_S / \mathfrak{o}_S^\times$, so the function H must possess complementary transformation properties for the sum to be well-defined.

Indeed, the function H satisfies a "twisted multiplicativity" in c, expressed in terms of n-th power residue symbols and depending on the root system Φ, which specializes to the usual multiplicativity when $n = 1$. Recall that the n-th power residue symbol $\left(\frac{c}{d}\right)_n$ is defined when c and d are coprime elements of $\mathfrak{o}_S$ and $\gcd(n, d) = 1$. It depends only on c modulo d, and satisfies the reciprocity law

$$\left(\frac{c}{d}\right)_n = (d, c)_{n,S} \left(\frac{d}{c}\right)_n.$$

(The properties of the power residue symbol and associated S-Hilbert symbols in our notation are set out in [12].) Then given $\boldsymbol{c} = (c_1, \cdots, c_r)$ and $\boldsymbol{c}' = (c_1', \cdots, c_r')$ in $\mathfrak{o}_S^r$ with $\gcd(c_1 \cdots c_r, c_1' \cdots c_r') = 1$, the function H satisfies

$$\frac{H(c_1 c_1', \cdots, c_r c_r'; \boldsymbol{m})}{H(\boldsymbol{c}; \boldsymbol{m}) \, H(\boldsymbol{c}'; \boldsymbol{m})} = \prod_{i=1}^{r} \left(\frac{c_i}{c_i'}\right)_n^{||\alpha_i||^2} \left(\frac{c_i'}{c_i}\right)_n^{||\alpha_i||^2} \prod_{i<j} \left(\frac{c_i}{c_j'}\right)_n^{2\langle \alpha_i, \alpha_j \rangle} \left(\frac{c_i'}{c_j}\right)_n^{2\langle \alpha_i, \alpha_j \rangle},$$

(1.3)

where α_i, $i = 1, \cdots, r$ denote the simple roots of Φ and we have chosen a Weyl group invariant inner product $\langle \cdot, \cdot \rangle$ for our root system embedded into a real vector space of dimension r. The inner product should be normalized so that for any $\alpha, \beta \in \Phi$, both $||\alpha||^2 = \langle \alpha, \alpha \rangle$ and $2\langle \alpha, \beta \rangle$ are integers. We will devote the majority of our attention to Φ of Type A, in which case we will assume the inner product is chosen so that all roots have length 1.

The function H possesses a further twisted multiplicativity with respect to the parameter $\boldsymbol{m}$. Given any

$$\boldsymbol{c} = (c_1, \cdots, c_r), \qquad \boldsymbol{m} = (m_1, \cdots, m_r), \qquad \boldsymbol{m}' = (m_1', \cdots, m_r')$$

with $\gcd(m_1' \cdots m_r', c_1 \cdots c_r) = 1$, H satisfies the twisted multiplicativity relation

$$H(\boldsymbol{c}; m_1 m_1', \cdots, m_r m_r') = \left(\frac{m_1'}{c_1}\right)_n^{-||\alpha_1||^2} \cdots \left(\frac{m_r'}{c_r}\right)_n^{-||\alpha_r||^2} H(\boldsymbol{c}; \boldsymbol{m}). \quad (1.4)$$

As a consequence of properties (1.3) and (1.4) the specification of H reduces to the case where the components of $\boldsymbol{c}$ and $\boldsymbol{m}$ are all powers of the same prime. Given a fixed prime p of $\mathfrak{o}_S$ and any $\boldsymbol{m} = (m_1, \cdots, m_r)$, let $l_i = \mathrm{ord}_p(m_i)$. Then we must specify $H(p^{k_1}, \cdots, p^{k_r}; p^{l_1}, \cdots, p^{l_r})$ for any r-tuple of nonnegative integers $\boldsymbol{k} = (k_1, \cdots, k_r)$. For brevity, we will refer to these coefficients as the "p-part" of H. To summarize, specifying a multiple Dirichlet series $Z_\Psi^{(n)}(\boldsymbol{s}; \boldsymbol{m}; \Phi)$ with chosen data is equivalent to specifying the p-parts of H.

Remark. Both the transformation property of Ψ in (1.1) and the definition of twisted multiplicativity in (1.3) depend on an enumeration of the simple roots of Φ. However, the product $H \cdot \Psi$ is independent of this enumeration of roots and furthermore well-defined modulo units, according to the reciprocity law. The p-parts of H are also independent of this enumeration of roots.

The definitions given above apply to any root system Φ. In most of this text, we will take Φ to be of Type A. In this case we will give two combinatorial definitions of the p-part of H. These two definitions of H will be referred to as H_Γ and H_Δ, and eventually shown to be equal. Thus either may be used to define the multiple Dirichlet series $Z(\boldsymbol{s}; \boldsymbol{m}; A_r)$. Both definitions will be given in terms of Gelfand-Tsetlin patterns.

By a *Gelfand-Tsetlin pattern of rank r* we mean an array of integers

$$
\mathfrak{T} = \left\{
\begin{array}{ccccccc}
a_{00} & & a_{01} & & a_{02} & \cdots & a_{0r} \\
& a_{11} & & a_{12} & & & a_{1r} \\
& & \ddots & & & \iddots & \\
& & & & a_{rr} & &
\end{array}
\right\} \tag{1.5}
$$

where the rows interleave; that is, $a_{i-1,j-1} \geqslant a_{i,j} \geqslant a_{i-1,j}$. Let $\lambda = (\lambda_1, \cdots, \lambda_r)$ be a dominant integral element for SL_{r+1}, so that $\lambda_1 \geqslant \lambda_2 \geqslant \cdots \geqslant \lambda_r$. In the next chapter, we will explain why Gelfand-Tsetlin patterns with top row $(\lambda_1, \cdots, \lambda_r, 0)$ are in bijection with basis vectors for the highest weight module for $SL_{r+1}(\mathbb{C})$ with highest weight λ.

The coefficients $H(p^{k_1}, \cdots, p^{k_r}; p^{l_1}, \cdots, p^{l_r})$ in both definitions H_Γ and H_Δ will be described in terms of Gelfand-Tsetlin patterns with top row (or equivalently, highest weight vector)

$$
\lambda + \rho = (l_1 + l_2 + \cdots + l_r + r, \cdots, l_{r-1} + l_r + 2, l_r + 1, 0). \tag{1.6}
$$

We denote by $\mathrm{GT}(\lambda + \rho)$ the set of all Gelfand-Tsetlin patterns having this top row. Here

$$
\rho = (r, r-1, \cdots, 0) \quad \text{and} \quad \lambda = (\lambda_1, \cdots, \lambda_{r+1}) \quad \text{where} \quad \lambda_i = \sum_{j \geqslant i} l_j. \tag{1.7}
$$

To any Gelfand-Tsetlin pattern $\mathfrak{T}$, we associate the following pair of functions with image in $\mathbb{Z}_{\geqslant 0}^r$:

$$
k_\Gamma(\mathfrak{T}) = (k_{\Gamma,1}(\mathfrak{T}), \cdots, k_{\Gamma,r}(\mathfrak{T})), \qquad k_\Delta(\mathfrak{T}) = (k_{\Delta,1}(\mathfrak{T}), \cdots, k_{\Delta,r}(\mathfrak{T})),
$$

where

$$
k_{\Gamma,i}(\mathfrak{T}) = \sum_{j=i}^{r} (a_{i,j} - a_{0,j}) \quad \text{and} \quad k_{\Delta,i}(\mathfrak{T}) = \sum_{j=r+1-i}^{r} (a_{0,j-r-1+i} - a_{r+1-i,j}). \tag{1.8}
$$

In the language of representation theory, the weight of the basis vector corresponding to the Gelfand-Tsetlin pattern $\mathfrak{T}$ can be read from differences of consecutive row sums in the pattern, so both k_Γ and k_Δ are expressions of the weight of the pattern up to an affine linear transformation.

Then given a fixed r-tuple of nonnegative integers $(l_1, \cdots, l_r)$, we make the following two definitions for p-parts of the multiple Dirichlet series:

$$
H_\Gamma(p^{k_1}, \cdots, p^{k_r}; p^{l_1}, \cdots, p^{l_r}) = \sum_{\substack{\mathfrak{T} \in \mathrm{GT}(\lambda+\rho) \\ k_\Gamma(\mathfrak{T}) = (k_1, \cdots, k_r)}} G_\Gamma(\mathfrak{T}) \tag{1.9}
$$

and

$$
H_\Delta(p^{k_1}, \cdots, p^{k_r}; p^{l_1}, \cdots, p^{l_r}) = \sum_{\substack{\mathfrak{T} \in \mathrm{GT}(\lambda+\rho) \\ k_\Delta(\mathfrak{T}) = (k_1, \cdots, k_r)}} G_\Delta(\mathfrak{T}), \tag{1.10}
$$

where the functions G_Γ and G_Δ on Gelfand-Tsetlin patterns will now be defined.

We will associate with $\mathfrak{T}$ two arrays $\Gamma(\mathfrak{T})$ and $\Delta(\mathfrak{T})$. The entries in these arrays are

$$\Gamma_{i,j} = \Gamma_{i,j}(\mathfrak{T}) = \sum_{k=j}^{r}(a_{i,k}-a_{i-1,k}), \qquad \Delta_{i,j} = \Delta_{i,j}(\mathfrak{T}) = \sum_{k=i}^{j}(a_{i-1,k-1}-a_{i,k}),$$

(1.11)

with $1 \leqslant i \leqslant j \leqslant r$, and we often think of attaching each entry of the array $\Gamma(\mathfrak{T})$ (resp. $\Delta(\mathfrak{T})$) with an entry of the pattern $a_{i,j}$ lying below the fixed top row. Thus we think of $\Gamma(\mathfrak{T})$ as applying a kind of *right-hand rule* to $\mathfrak{T}$, since $\Gamma_{i,j}$ involves entries above and to the right of $a_{i,j}$ as in (1.11); in Δ we use a *left-hand rule* where $\Delta_{i,j}$ involves entries above and to the left of $a_{i,j}$ as in (1.11). When we represent these arrays graphically, we will right-justify the Γ array and left-justify the Δ array. For example, if

$$\mathfrak{T} = \left\{ \begin{array}{cccc} 12 & 9 & 4 & 0 \\ & 10 & 5 & 3 \\ & & 7 & 4 \\ & & & 6 \end{array} \right\}$$

then

$$\Gamma(\mathfrak{T}) = \begin{bmatrix} 5 & 4 & 3 \\ & 3 & 1 \\ & & 2 \end{bmatrix} \quad \text{and} \quad \Delta(\mathfrak{T}) = \begin{bmatrix} 2 & 6 & 7 \\ 3 & 4 & \\ 1 & & \end{bmatrix}.$$

To provide the definitions of G_Γ and G_Δ corresponding to each array, it is convenient to *decorate* the entries of the Γ and Δ arrays by boxing or circling certain of them. Using the *right-hand rule* with the Γ array, if $a_{i,j} = a_{i-1,j-1}$ then we say $\Gamma_{i,j}$ is *boxed*, and indicate this when we write the array by putting a box around it, while if $a_{i,j} = a_{i-1,j}$ we say it is *circled* (and we circle it). Using the *left-hand rule* to obtain the Δ array, we box $\Delta_{i,j}$ if $a_{i,j} = a_{i-1,j}$ and we circle it if $a_{i,j} = a_{i-1,j-1}$. For example, if

$$\mathfrak{T} = \left\{ \begin{array}{cccc} 12 & 10 & 4 & 0 \\ & 10 & 5 & 3 \\ & & 7 & 5 \\ & & & 6 \end{array} \right\} \tag{1.12}$$

then the decorated arrays are

$$\Gamma(\mathfrak{T}) = \begin{bmatrix} \textcircled{4} & 4 & 3 \\ & 4 & \boxed{2} \\ & & 1 \end{bmatrix}, \quad \Delta(\mathfrak{T}) = \begin{bmatrix} \boxed{2} & 7 & 8 \\ 3 & \textcircled{3} & \\ 1 & & \end{bmatrix}. \tag{1.13}$$

We sometimes use the terms *right-hand rule* and *left-hand rule* to refer to both the direction of accumulation of the row differences, and to the convention for decorating these accumulated differences.

If $m, c \in \mathfrak{o}_S$ with $c \neq 0$ define the Gauss sum

$$g(m,c) = \sum_{a \bmod c} \left(\frac{a}{c}\right)_n \psi\left(\frac{am}{c}\right), \tag{1.14}$$

where ψ is a character of F_S that is trivial on $\mathfrak{o}_S$ and no larger fractional ideal. With p now fixed, for brevity let

$$g(a) = g(p^{a-1}, p^a) \quad \text{and} \quad h(a) = g(p^a, p^a). \tag{1.15}$$

These functions will only occur with $a > 0$. The reader may check that $g(a)$ is nonzero for any value of a, while $h(a)$ is nonzero only if $n|a$, in which case $h(a) = (q-1)q^{a-1}$, where q is the cardinality of $\mathfrak{o}_S/p\mathfrak{o}_S$. Thus if $n|a$ then $h(a) = \phi(p^a)$, the Euler phi function for $p^a\mathfrak{o}_S$.

Let

$$G_\Gamma(\mathfrak{T}) = \prod_{1 \leqslant i \leqslant j \leqslant r} \begin{cases} g(\Gamma_{ij}) & \text{if } \Gamma_{ij} \text{ is boxed but not circled in } \Gamma(\mathfrak{T}); \\ q^{\Gamma_{ij}} & \text{if } \Gamma_{ij} \text{ is circled but not boxed;} \\ h(\Gamma_{ij}) & \text{if } \Gamma_{ij} \text{ is neither circled nor boxed;} \\ 0 & \text{if } \Gamma_{ij} \text{ is both circled and boxed.} \end{cases}$$

We say that the pattern $\mathfrak{T}$ is *strict* if $a_{i,j} > a_{i,j+1}$ for every $0 \leqslant i \leqslant j < r$. Thus the rows of the pattern are a strictly decreasing sequence. Otherwise we say that the pattern is *non-strict*. It is clear from the definitions that $\mathfrak{T}$ is non-strict if and only if $\Gamma(\mathfrak{T})$ has an entry that is both boxed and circled, so $G_\Gamma(\mathfrak{T}) = 0$ for non-strict patterns. Similarly let

$$G_\Delta(\mathfrak{T}) = \prod_{1 \leqslant i \leqslant j \leqslant r} \begin{cases} g(\Delta_{ij}) & \text{if } \Delta_{ij} \text{ is boxed but not circled in } \Delta(\mathfrak{T}); \\ q^{\Delta_{ij}} & \text{if } \Delta_{ij} \text{ is circled but not boxed;} \\ h(\Delta_{ij}) & \text{if } \Delta_{ij} \text{ is neither circled nor boxed;} \\ 0 & \text{if } \Delta_{ij} \text{ is both circled and boxed.} \end{cases}$$

For example, we may use the decorated arrays as in (1.13) to write down $G_\Gamma(\mathfrak{T})$ and $G_\Delta(\mathfrak{T})$ for the pattern $\mathfrak{T}$ appearing in (1.12) as follows:

$$G_\Gamma(\mathfrak{T}) = q^4 h(4)h(3)h(4)g(2)h(1), \quad \text{and} \quad G_\Delta(\mathfrak{T}) = g(2)h(7)h(8)h(3)q^3 h(1).$$

Inserting the respective definitions for G_Γ and G_Δ into the formulas (1.9) and (1.10) completes the two definitions of the p-parts of H_Γ and H_Δ, and with it two definitions for a multiple Dirichlet series $Z_\Psi(s; m)$. For example, the pattern $\mathfrak{T}$ in (1.12) would appear in the p-part of $Z_\Psi(s; m)$ if

$$(\text{ord}_p(m_1), \cdots, \text{ord}_p(m_r)) = (1, 5, 3)$$

according to the top row of $\mathfrak{T}$. Moreover, this pattern would contribute $G_\Gamma(\mathfrak{T})$ to the term $H_\Gamma(p^k; p^l) = H_\Gamma(p^4, p^8, p^6; p^1, p^5, p^3)$ according to the definitions in (1.8) and (1.9).

In [17], the definition H_Γ was used to define the series, and so we will state our theorem on functional equations and analytic continuation of $Z_\Psi(s; m)$ using this choice. Before stating the result precisely, we need to define certain normalizing factors for the multiple Dirichlet series. These have a uniform description for all root systems (see Section 3.3 of [12]), but for simplicity we state them only for Type A here.

Let

$$\boldsymbol{G}_n(s) = (2\pi)^{-2(n-1)s} n^{2ns} \prod_{j=1}^{n-1} \Gamma\left(2s - 1 + \frac{j}{n}\right). \tag{1.16}$$

We will identify the weight space for $GL(r+1, \mathbb{C})$ with $\mathbb{Z}^{r+1}$ in the usual way. For any $\alpha \in \Phi^+$, the set of positive roots, there exist $1 \leqslant i < j \leqslant r+1$ such that $\alpha = \alpha_{i,j}$ is the root $(0, \cdots, 0, 1, 0, \cdots, -1, 0, \cdots)$ with the 1 in the i-th place and the -1 in the j-th place. If $\alpha = \alpha_{i,j}$ is a positive root, then define

$$G_\alpha(s) = G_n \left(\frac{1}{2} + \left(s_i + s_{i+1} + \cdots + s_{j-1} - \frac{j-i}{2} \right) \right). \tag{1.17}$$

Further let

$$\zeta_\alpha(s) = \zeta \left(1 + 2n \left(s_i + s_{i+1} + \cdots + s_{j-1} - \frac{j-i}{2} \right) \right)$$

where ζ is the Dedekind zeta function attached to the number field F. Then the normalized multiple Dirichlet series is given by

$$Z_\Psi^*(s; m) = \left[\prod_{\alpha \in \Phi^+} G_\alpha(s) \zeta_\alpha(s) \right] Z_\Psi(s, m). \tag{1.18}$$

THEOREM 1.1 *The Weyl group multiple Dirichlet series $Z_\Psi^*(s; m)$ with coefficients H_Γ as in (1.9) has meromorphic continuation to $\mathbb{C}^r$ and satisfies functional equations*

$$Z_\Psi^*(s; m) = |m_i|^{1-2s_i} Z_{\sigma_i \Psi}^*(\sigma_i s; m) \tag{1.19}$$

for all simple reflections $\sigma_i \in W$, where

$$\sigma_i(s_i) = 1 - s_i, \qquad \sigma_i(s_j) = \begin{cases} s_i + s_j - 1/2 & \text{if } i, j \text{ adjacent,} \\ s_j & \text{otherwise.} \end{cases}$$

Here $\sigma_i : \mathcal{M} \to \mathcal{M}$ is a linear map defined in [14].

The endomorphisms σ_i of the space $\mathcal{M}$ of functions satisfying (1.1) are the simple reflections in an action of the Weyl group W on $\mathcal{M}$. See [12] and [14] for more information.

This proves Conjecture 2 of [17]. An explicit description of the polar hyperplanes of Z_Ψ^* can be found in Section 7 of [14]. As we demonstrate in Chapter 6, this theorem ultimately follows from proving the equivalence of the two definitions of the p-part H_Γ and H_Δ offered in (1.9) and (1.10). Because of this implication, and because it is of interest to construct such functions attached to a representation but independent of choices of coordinates (a notion we make precise in subsequent chapters using the crystal description), we consider the equivalence of these two descriptions to be our main theorem.

THEOREM 1.2 *We have $H_\Gamma = H_\Delta$.*

The proof is outlined in Chapter 6 and completed in the subsequent chapters.

To give a flavor for this result, we give one example. Suppose that $r = 2$, $(k_1, k_2) = (6, 6)$, and $(l_1, l_2) = (2, 5)$. Then the following strict Gelfand-Tsetlin patterns contribute to $H_\Gamma(p^6, p^6; p^2, p^5)$:

$$\left\{ \begin{matrix} 9 && 6 && 0 \\ & 9 && 3 & \\ && 6 && \end{matrix} \right\}, \quad \left\{ \begin{matrix} 9 && 6 && 0 \\ & 8 && 4 & \\ && 6 && \end{matrix} \right\}, \quad \left\{ \begin{matrix} 9 && 6 && 0 \\ & 7 && 5 & \\ && 6 && \end{matrix} \right\}.$$

These give
$$H_\Gamma(p^6, p^6; p^2, p^5) = g(6)h(3)^2 + h(2)h(4)h(6) + h(1)h(5)h(6). \qquad (1.20)$$

Similarly the following Gelfand-Tsetlin patterns contribute to H_Δ:

$$\left\{ \begin{matrix} 9 & & 6 & & 0 \\ & 9 & & 0 & \\ & & 3 & & \end{matrix} \right\}, \left\{ \begin{matrix} 9 & & 6 & & 0 \\ & 8 & & 1 & \\ & & 3 & & \end{matrix} \right\},$$

$$\left\{ \begin{matrix} 9 & & 6 & & 0 \\ & 7 & & 2 & \\ & & 3 & & \end{matrix} \right\}, \left\{ \begin{matrix} 9 & & 6 & & 0 \\ & 6 & & 3 & \\ & & 3 & & \end{matrix} \right\}.$$

These give
$$H_\Delta(p^6, p^6; p^2, p^5) = g(6)h(6) + h(1)h(5)h(6) + h(2)h(4)h(6) + g(3)^2 h(6). \qquad (1.21)$$

In this special case, two terms in the expressions for H_Γ and H_Δ match exactly, a point that we will return to in (1.22). Theorem 1.2 is true for all n, since

$$g(6)h(3)^2 = g(6)h(6) + g(3)^2 h(6).$$

Although this identity is true for all n, its meaning is different for different values of n. Indeed, both sides are zero unless n divides 6. When $n = 2$ or 6, the left-hand side vanishes since $h(3) = 0$, while the right-hand side vanishes since, by elementary properties of Gauss sums (see Proposition 8.1), $g(6) = -q^5$ and $g(3)^2 = q^5$. When $n = 1$ or $n = 3$, both sides are nonzero, and the equality is the identity

$$-q^5(q^3 - q^2)^2 = -q^5(q^6 - q^5) + (-q^2)^2(q^6 - q^5).$$

The identity of Theorem 1.2 equates two sums. The example that we have just given shows that it is not possible to give a term-by-term comparison of the two sums (1.9) and (1.10). That is, there is no bijection $\mathfrak{T} \mapsto \mathfrak{T}'$ between the set of Gelfand-Tsetlin patterns with top row $\lambda + \rho$ and itself such that $k_\Gamma(\mathfrak{T}) = k_\Delta(\mathfrak{T}')$ and $G_\Gamma(\mathfrak{T}) = G_\Delta(\mathfrak{T}')$. Still there is a bijection such that $k_\Gamma(\mathfrak{T}) = k_\Delta(\mathfrak{T}')$ and such that the identity $G_\Gamma(\mathfrak{T}) = G_\Delta(\mathfrak{T}')$ is *often* true. In other words, many terms in (1.9) are equal to corresponding terms in (1.10).

The involution on the set of Gelfand-Tsetlin patterns is called the *Schützenberger involution*. It was originally introduced by Schützenberger [66] in the context of tableaux. The involution was transported to the setting of Gelfand-Tsetlin patterns by Kirillov and Berenstein [51], and defined for general crystals (to be discussed in Chapter 2) by Lusztig [58]. We give its definition now.

Given a Gelfand-Tsetlin pattern (1.5), the condition that the rows interleave means that each $a_{i,j}$ is constrained by the inequalities

$$\min(a_{i-1,j-1}, a_{i+1,j}) \leqslant a_{i,j} \leqslant \max(a_{i-1,j}, a_{i+1,j+1}).$$

This means that we can reflect the entry $a_{i,j}$ across the midpoint of this interval and obtain another Gelfand-Tsetlin pattern. Thus we replace every entry $a_{i,j}$ in the i-th row by

$$a'_{i,j} = \min(a_{i-1,j-1}, a_{i+1,j}) + \max(a_{i-1,j}, a_{i+1,j+1}) - a_{i,j}.$$

This requires interpretation if $j = i$ or $j = r$. For these cases, we will set

$$a'_{i,i} = a_{i-1,i-1} + \max(a_{i-1,i}, a_{i+1,i+1}) - a_{i,i}$$

and

$$a'_{i,r} = \min(a_{i-1,r-1}, a_{i+1,r}) + a_{i-1,r} - a_{i,r},$$

unless $i = j = r$, and then we set $a'_{r,r} = a_{r-1,r-1} + a_{r-1,r} - a_{r,r}$. This operation on the entire row will be denoted by t_{r+1-i}. Note that it only affects this lone row in the pattern. Further involutions on patterns may be built out of the t_i, and will be called q_i following Berenstein and Kirillov. Let q_0 be the identity map, and define recursively $q_i = t_1 t_2 \cdots t_i q_{i-1}$. The t_i have order two. They do not satisfy the braid relation, so $t_i t_{i+1} t_i \neq t_{i+1} t_i t_{i+1}$. However $t_i t_j = t_j t_i$ if $|i - j| > 1$ and this implies that the q_i also have order two. The operation q_r is the *Schützenberger involution* that we mentioned earlier.

For example, let $r = 2$, and let us compute q_2 of a typical Gelfand-Tsetlin pattern. Following the algorithm outlined above, if

$$\mathfrak{T} = \left\{ \begin{array}{ccccc} 9 & & 6 & & 0 \\ & 7 & & 5 & \\ & & 6 & & \end{array} \right\},$$

then

$$q_2(\mathfrak{T}) = \left\{ \begin{array}{ccccc} 9 & & 6 & & 0 \\ & 8 & & 1 & \\ & & 3 & & \end{array} \right\}.$$

Indeed, $q_2 = t_1 t_2 t_1$ and we compute:

$$\left\{ \begin{array}{ccccc} 9 & & 6 & & 0 \\ & 7 & & 5 & \\ & & 6 & & \end{array} \right\} \xrightarrow{t_1} \left\{ \begin{array}{ccccc} 9 & & 6 & & 0 \\ & 7 & & 5 & \\ & & 6 & & \end{array} \right\} \xrightarrow{t_2}$$

$$\left\{ \begin{array}{ccccc} 9 & & 6 & & 0 \\ & 8 & & 1 & \\ & & 6 & & \end{array} \right\} \xrightarrow{t_1} \left\{ \begin{array}{ccccc} 9 & & 6 & & 0 \\ & 8 & & 1 & \\ & & 3 & & \end{array} \right\}.$$

Now observe that

$$G_\Gamma(\mathfrak{T}) = h(6)\, h(5)\, h(1) = G_\Delta(q_2 \mathfrak{T}). \tag{1.22}$$

This accounts for the equality of one of the terms in the sum (1.20) and one of the terms in the sum (1.21), and illustrates the point that *often* $G_\Gamma(\mathfrak{T}) = G_\Delta(q_r \mathfrak{T})$. The difficulty of Theorem 1.2 is the problem of accounting for the exceptions to this rule in a systematic way.

It is sometimes useful to modify the coefficients in the Dirichlet series as follows. If a is a positive integer let

$$g^\flat(a) = q^{-a} g(a), \qquad h^\flat(a) = q^{-a} h(a). \tag{1.23}$$

These "reduced" coefficients have the property that they depend only on a modulo n. Let

$$G_\Gamma^\flat(\mathfrak{T}) = \prod_{1 \leqslant i \leqslant j \leqslant r} \left\{ \begin{array}{ll} g^\flat(\Gamma_{ij}) & \text{if } \Gamma_{ij} \text{ is boxed but not circled in } \Gamma(\mathfrak{T}); \\ 1 & \text{if } \Gamma_{ij} \text{ is circled but not boxed;} \\ h^\flat(\Gamma_{ij}) & \text{if } \Gamma_{ij} \text{ is neither circled nor boxed;} \\ 0 & \text{if } \Gamma_{ij} \text{ is both circled and boxed,} \end{array} \right.$$

and similarly define

$$G_\Delta^\flat(\mathfrak{T}) = \prod_{1 \leqslant i \leqslant j \leqslant r} \begin{cases} g^\flat(\Delta_{ij}) & \text{if } \Delta_{ij} \text{ is boxed but not circled in } \Delta(\mathfrak{T}); \\ 1 & \text{if } \Delta_{ij} \text{ is circled but not boxed}; \\ h^\flat(\Delta_{ij}) & \text{if } \Delta_{ij} \text{ is neither circled nor boxed}; \\ 0 & \text{if } \Delta_{ij} \text{ is both circled and boxed}. \end{cases}$$

It is not hard to check that

$$G_\Gamma(\mathfrak{T}) = q^{\sum_{i=1}^r k_{\Gamma,i}(\mathfrak{T})} G_\Gamma^\flat(\mathfrak{T}), \qquad G_\Delta(\mathfrak{T}) = q^{\sum_{i=1}^r k_{\Delta,i}(\mathfrak{T})} G_\Delta^\flat(\mathfrak{T}).$$

It follows that if we define

$$H_\Gamma^\flat(p^{k_1}, \cdots, p^{k_r}; p^{l_1}, \cdots, p^{l_r}) = \sum_{k_\Gamma(\mathfrak{T})=(k_1,\cdots,k_r)} G_\Gamma^\flat(\mathfrak{T}) \qquad (1.24)$$

and similarly let

$$H_\Delta^\flat(p^{k_1}, \cdots, p^{k_r}; p^{l_1}, \cdots, p^{l_r}) = \sum_{k_\Delta(\mathfrak{T})=(k_1,\cdots,k_r)} G_\Delta^\flat(\mathfrak{T}), \qquad (1.25)$$

then extend these reduced coefficients $H^\flat$ by the same multiplicativities (1.3) and (1.4) as H, we have

$$Z_\Psi(\boldsymbol{s}; \boldsymbol{m}) = \sum_{\substack{\boldsymbol{c}=(c_1,\cdots,c_r)\in(\mathfrak{o}_S/\mathfrak{o}_S^\times)^r \\ c_i \neq 0}} \frac{H^\flat(\boldsymbol{c}; \boldsymbol{m})}{\mathbb{N}c_1^{2s_1-1}\cdots\mathbb{N}c_r^{2s_r-1}}.$$

Chapter Two

Crystals and Gelfand-Tsetlin Patterns

We will translate the definitions of the Γ and Δ arrays in (1.11), and hence of the multiple Dirichlet series, into the language of crystal bases. The entries in these arrays and the accompanying boxing and circling rules will be reinterpreted in terms of the Kashiwara operators. Thus, what appeared as a pair of unmotivated functions on Gelfand-Tsetlin patterns in the previous chapter now takes on intrinsic representation theoretic meaning. Despite the conceptual importance of this reformulation, the reader can skip this chapter and the subsequent chapters devoted to crystals with no loss of continuity. For further background information on crystals, we recommend Hong and Kang [42] and Kashiwara [47] as basic references. After completing the proof of our main theorem, the subject of crystals is taken up again in Chapter 18, where we reformulate aspects of the proof in terms of crystals. Then in Chapter 20 we explain the notion of crystal bases more generally and explain why they appear naturally in the description of p-parts.

In the present chapter, we restrict to crystals of Cartan type A_r and identify the weight lattice Λ of $\mathfrak{gl}_{r+1}(\mathbb{C})$ with $\mathbb{Z}^{r+1}$. The weight $\lambda = (\lambda_1, \cdots, \lambda_{r+1}) \in \mathbb{Z}^{r+1}$ is *dominant* if $\lambda_1 \geqslant \lambda_2 \geqslant \ldots$. If $\lambda_{r+1} \geqslant 0$ we call the dominant weight *effective*. Thus an effective dominant weight is just a partition of length $\leqslant r + 1$. Let $e_i = (0, \cdots, 1, \cdots, 0) \in \mathbb{Z}^{r+1}$ where the 1 is in the i-th position. Then the root system Φ of Type A consists of the vectors

$$e_i - e_j \in \Lambda$$

with $i \neq j$. The positive roots Φ^+ consist of the roots $e_i - e_j$ with $i < j$. We will denote by ρ the *Weyl vector*, already defined in (1.7) to be $(r, r - 1, \cdots, 2, 1, 0)$. Regarding ρ as an element of the weight lattice, it is not half the sum of the positive roots, but

$$\rho - \frac{1}{2} \sum_{\alpha \in \Phi^+} \alpha$$

is orthogonal to the roots. This means that we may use ρ and $\frac{1}{2} \sum_{\alpha \in \Phi^+} \alpha$ interchangeably in places such as the Weyl character formula as written in Chapter 5.

If λ is a dominant weight then there is a *crystal graph* $\mathcal{B}_\lambda$ with highest weight λ. It is equipped with a *weight function* $\mathrm{wt} : \mathcal{B}_\lambda \longrightarrow \mathbb{Z}^{r+1}$ such that if μ is any weight and if $m(\mu, \lambda)$ is the multiplicity of μ in the irreducible representation of $\mathrm{GL}_{r+1}(\mathbb{C})$ with highest weight λ then $m(\mu, \lambda)$ is also the number of $v \in \mathcal{B}_\lambda$ with $\mathrm{wt}(v) = \mu$. It has operators

$$e_i, f_i : \mathcal{B}_\lambda \longrightarrow \mathcal{B}_\lambda \cup \{0\} \quad (1 \leqslant i \leqslant r)$$

such that:

- if $e_i(v) \neq 0$ then $v = f_i(e_i(v))$ and $\mathrm{wt}(e_i(v)) = \mathrm{wt}(v) + \alpha_i$,

- if $f_i(v) \neq 0$ then $e_i(f_i(v)) = v$ and $\mathrm{wt}(f_i(v)) = \mathrm{wt}(v) - \alpha_i$.

Here $\alpha_1 = (1, -1, 0, \cdots, 0)$, $\alpha_2 = (0, 1, -1, 0, \cdots, 0)$, etc. are the simple roots in the usual order. These root operators give $\mathcal{B}_\lambda$ the structure of a directed graph with edges labeled from the set $1, 2, \cdots, r$. The vertices v and w are connected by an edge $v \xrightarrow{i} w$ or $v \xrightarrow{f_i} w$ if $w = f_i(v)$.

Remark. The elements of $\mathcal{B}_\lambda$ are basis vectors for a representation of the quantized enveloping algebra $U_q(\mathfrak{gl}_{r+1}(\mathbb{C}))$. Strictly speaking we should reserve f_i for the root operators in this quantized enveloping algebra and so distinguish between f_i and $\tilde{f}_i$, the operators used to define the crystal basis, as in [48]. However we will not actually use the quantum group but only the crystal graph, so we will simplify the notation by writing f_i instead of $\tilde{f}_i$, and similarly for the e_i. We will use the terms *crystal*, *crystal base* and *crystal graph* interchangeably.

The crystal graph $\mathcal{B}_\lambda$ has an involution $\mathrm{Sch} : \mathcal{B}_\lambda \longrightarrow \mathcal{B}_\lambda$ that such that

$$\mathrm{Sch} \circ e_i = f_{r+1-i} \circ \mathrm{Sch}, \qquad \mathrm{Sch} \circ f_i = e_{r+1-i} \circ \mathrm{Sch}. \qquad (2.1)$$

In addition to the involution Sch there is a bijection $\psi_\lambda : \mathcal{B}_\lambda \longrightarrow \mathcal{B}_{-w_0\lambda}$ such that

$$\psi_\lambda \circ f_i = e_i \circ \psi_\lambda, \qquad \psi_\lambda \circ e_i = f_i \circ \psi_\lambda. \qquad (2.2)$$

Here w_0 is the long Weyl group element. If $\lambda = (\lambda_1, \cdots, \lambda_{r+1})$ is a dominant weight then $-w_0\lambda = (-\lambda_{r+1}, \cdots, -\lambda_1)$ is also a dominant weight so there is a crystal $\mathcal{B}_{-w_0\lambda}$ with that highest weight. The map ψ_λ commutes with Sch and the composition $\phi_\lambda = \mathrm{Sch} \circ \psi_\lambda = \psi_\lambda \circ \mathrm{Sch}$ has the effect

$$\phi_\lambda \circ f_i = f_{r+1-i} \circ \phi_\lambda, \qquad \phi_\lambda \circ e_{r+1-i} = e_i \circ \phi_\lambda. \qquad (2.3)$$

The involution Sch was first described by Schützenberger [66] in the context of tableaux. It was transported to the setting of Gelfand-Tsetlin patterns by Kirillov and Berenstein [51], and defined for general crystals by Lusztig [58]. Another useful reference for the involutions is Lenart [55]. We will demonstrate the equivalence of Sch on $\mathcal{B}_\lambda$ and the involution q_r on Gelfand-Tsetlin patterns of Chapter 1 in Proposition 2.2 below. First we must explain the bijection between Type A crystals $\mathcal{B}_\lambda$ and Gelfand-Tsetlin patterns.

If we remove all edges labeled r from the crystal graph $\mathcal{B}_\lambda$, then we obtain a crystal graph of rank $r - 1$. It inherits a weight function from $\mathcal{B}_\lambda$, which we compose with the projection $\mathbb{Z}^{r+1} \longrightarrow \mathbb{Z}^r$ onto the first r coordinates.

The restricted crystal may be disconnected, in which case it is a disjoint union of crystals of Cartan type A_{r-1}, and the crystals that appear in this restriction are described by the following *branching rule*:

$$\mathcal{B}_\lambda = \bigcup_{\substack{\mu = (\mu_1, \cdots, \mu_r) \\ \mu \text{ dominant} \\ \lambda, \mu \text{ interleave}}} \mathcal{B}_\mu \qquad (2.4)$$

where the "interleave" condition means that μ runs through dominant weights such that

$$\lambda_1 \geqslant \mu_1 \geqslant \lambda_2 \geqslant \mu_2 \geqslant \ldots \geqslant \lambda_{r+1}.$$

The branching rule is multiplicity-free, meaning that no crystal $\mathcal{B}_\mu$ occurs more than once on the right-hand side of (2.4). Since representations of the quantized enveloping algebra $U_q\big(\mathfrak{gl}_{r+1}(\mathbb{C})\big)$ correspond to representations of $\mathrm{GL}_{r+1}(\mathbb{C})$, the decomposition (2.4) follows from the well-known branching rule from $\mathrm{GL}_{r+1}(\mathbb{C})$ to $\mathrm{GL}_r(\mathbb{C})$ (Pieri's rule). See, for example, Bump [19], Chapter 44.

The bijection between the crystal graph with highest weight λ and Gelfand-Tsetlin patterns with top row λ may be seen from the branching rule described above. Given $v \in \mathcal{B}_\lambda$, we first branch down from A_r to A_{r-1}, selecting the unique crystal $\mathcal{B}_\mu$ from (2.4) with $v \in \mathcal{B}_\mu$, that is, the connected component of the restricted crystal that contains v. Then λ and μ are the first two rows of the Gelfand-Tsetlin pattern. Continuing to branch down to $A_{r-2}, A_{r-3}, \cdots$ we may read off the remaining rows of the pattern. Let $\mathfrak{T}_v$ be the resulting Gelfand-Tsetlin pattern.

The crystal $\mathcal{B}_\lambda$ contains $\mathcal{B}_\mu$ if and only if λ and μ interleave, which is equivalent to $-w_0\lambda$ and $-w_0\mu$ interleaving, and hence if and only if $\mathcal{B}_{-w_0\lambda}$ contains $\mathcal{B}_{-w_0\mu}$. The operation ψ_λ in (2.2) that reverses the root operators must be compatible with this branching rule, and so each row of $\mathfrak{T}_{\psi_\lambda v}$ is obtained from the corresponding row of $\mathfrak{T}_v$ by reversing the entries and changing their sign. Thus, denoting by "rev" the operation of reversing an array from left to right and by $-\mathfrak{T}$ the pattern with all entries negated, we have

$$\mathfrak{T}_{\psi_\lambda v} = -\mathfrak{T}_v^{\mathrm{rev}}. \tag{2.5}$$

An alternative way of getting this bijection comes from the interpretation of crystals as crystals of tableaux. We will assume that λ is effective, that is, that its entries are nonnegative. Then Gelfand-Tsetlin patterns with top row λ are in bijection with semistandard Young tableaux with shape λ and labels in $\{1, 2, 3, \cdots, r+1\}$. In this bijection, one starts with a tableau, and successively reduces to a series of smaller tableaux by eliminating the entries. Thus if $r + 1 = 4$, starting with the tableau

$$\mathcal{T} = \begin{array}{|c|c|c|c|} \hline 1 & 1 & 2 & 2 \\ \hline 2 & 4 \\ \cline{1-2} 3 \\ \cline{1-1} \end{array}$$

and eliminating $4, 3, 2, 1$ successively one has the following sequence of tableaux:

$$\begin{array}{|c|c|c|c|} \hline 1 & 1 & 2 & 2 \\ \hline 2 & 4 \\ \cline{1-2} 3 \\ \cline{1-1} \end{array} \longrightarrow \begin{array}{|c|c|c|c|} \hline 1 & 1 & 2 & 2 \\ \hline 2 \\ \cline{1-1} 3 \\ \cline{1-1} \end{array} \longrightarrow \begin{array}{|c|c|c|c|} \hline 1 & 1 & 2 & 2 \\ \hline 2 \\ \cline{1-1} \end{array} \longrightarrow \begin{array}{|c|c|} \hline 1 & 1 \\ \hline \end{array} \quad .$$

Now reading off the shapes of these tableaux gives $r + 1$ shapes that are the rows of a Gelfand-Tsetlin pattern $\mathfrak{T}$. In this example:

$$\mathcal{T} \mapsto \mathfrak{T} = \left\{ \begin{array}{ccccccc} 4 & & 2 & & 1 & & 0 \\ & 4 & & 1 & & 1 & \\ & & 4 & & 1 & & \\ & & & 2 & & & \end{array} \right\}.$$

In discussing the bijection between Gelfand-Tsetlin patterns and tableaux, we have assumed that λ is effective, but what if it is not? If λ is a dominant weight, so is

$$\lambda + (m^{r+1}) = (\lambda_1 + m, \cdots, \lambda_{r+1} + m)$$

for any m. (As usual, $(m^k) = (m, \cdots, m)$ is the partition with k parts each equal to m.) We will denote the corresponding crystal $\mathcal{B}_{\lambda+(m^{r+1})} = \det^m \otimes \mathcal{B}_\lambda$ since this operation corresponds to tensoring with a power of the determinant character for representations of $\mathfrak{gl}_{r+1}(\mathbb{C})$. There is a bijection from $\mathcal{B}_\lambda$ to $\det^m \otimes \mathcal{B}_\lambda$ that is compatible with the root operators and shifts the weight by (m^{r+1}). If λ is not effective, still $\lambda + (m^{r+1})$ is effective for sufficiently large m. On the other hand, if λ *is* effective (so there is a bijection with tableaux of shape λ) then it is instructive to consider the effect of this operation on tableaux corresponding to the bijection $\mathcal{B}_\lambda \longrightarrow \det^m \otimes \mathcal{B}_\lambda$. It simply adds m columns of the form

1
2
$\vdots$
$r+1$

at the beginning of the tableau. So if λ is not effective, we may still think of $\mathcal{B}_\lambda$ as being in bijection with a crystal of tableaux with the weight operator shifted by (n^{r+1}), which amounts to "borrowing" n columns of this form.

Returning to the effective case, the tableau $\mathcal{T}$ parametrizes a vector in a tensor power of the standard module of the quantum group $U_q(\mathfrak{sl}_{r+1}(\mathbb{C}))$ as follows. Following the notations in Kashiwara and Nakashima [48] the *standard crystal* (corresponding to the standard representation) has basis $\boxed{i}$ $(i = 1, 2, \cdots, r+1)$. The highest weight vector is $\boxed{1}$ and the root operators have the effect $\boxed{i} \xrightarrow{f_i} \boxed{i+1}$.

The tensor product operation on crystals is described in Kashiwara and Nakashima [48], or in Hong and Kang [42]. If $\mathcal{B}$ and $\mathcal{B}'$ are crystals, then $\mathcal{B} \otimes \mathcal{B}'$ consists of all pairs $x \otimes y$ with $x \in \mathcal{B}$ and $y \in \mathcal{B}'$. The root operators have the following effect:

$$f_i(x \otimes y) = \begin{cases} f_i(x) \otimes y & \text{if } \phi_i(x) > \varepsilon_i(y), \\ x \otimes f_i(y) & \text{if } \phi_i(x) \leqslant \varepsilon_i(y), \end{cases}$$
$$e_i(x \otimes y) = \begin{cases} e_i(x) \otimes y & \text{if } \phi_i(x) \geqslant \varepsilon_i(y), \\ x \otimes e_i(y) & \text{if } \phi_i(x) < \varepsilon_i(y). \end{cases}$$

Here $\phi_i(x)$ is the largest integer ϕ such that $f_i^\phi(x) \neq 0$ and similarly $\varepsilon_i(x)$ is the largest integer ε such that $e_i^\varepsilon(x) \neq 0$.

Now tableaux are turned into elements of a tensor power of the standard crystal by reading the columns from top to bottom, and taking the columns in order from right to left. Thus the tableau

1	1	2	2
2	4		
3			

becomes $\boxed{2} \otimes \boxed{2} \otimes \boxed{1} \otimes \boxed{4} \otimes \boxed{1} \otimes \boxed{2} \otimes \boxed{3}$.

The set of vectors coming from tableaux with shape λ form a subcrystal of the tensor power of the standard crystal. This crystal of tableaux has highest weight λ and is isomorphic to $\mathcal{B}_\lambda$. Thus we get bijections

$$\mathcal{B}_\lambda \longleftrightarrow \left\{ \begin{array}{c} \text{Tableaux in } 1, \cdots, r \\ \text{with shape } \lambda \end{array} \right\} \longleftrightarrow \left\{ \begin{array}{c} \text{Gelfand-Tsetlin patterns} \\ \text{with top row } \lambda \end{array} \right\}. \tag{2.6}$$

This is the same as the bijection between $\mathcal{B}_\lambda$ and Gelfand-Tsetlin patterns that was described previously in terms of branching rules. Indeed, the branching rule for tableaux is as follows. Beginning with a tableau $\mathcal{T}$ in $1, \cdots, r$ of shape λ, erase all r's. This produces a tableau $\mathcal{T}'$ of shape μ where λ and μ interleave, and the Gelfand-Tsetlin pattern of $\mathcal{T}'$ is the Gelfand-Tsetlin pattern corresponding to $\mathcal{T}$ minus its top row.

We will soon explain yet another way of relating the Gelfand-Tsetlin pattern to $v \in \mathcal{B}_\lambda$. This is based on ideas in Lusztig [59], Berenstein and Zelevinsky [5, 4] and Littelmann [57]. Let w be an element of the Weyl group W, and let us give a reduced decomposition of w into simple reflections. That is, if $l(w)$ is the length of w, let $1 \leqslant \Omega_i \leqslant r$ be given $(1 \leqslant i \leqslant l(w))$ such that

$$w = \sigma_{\Omega_1} \sigma_{\Omega_2} \cdots \sigma_{\Omega_N}.$$

We call the sequence $\Omega_1, \cdots, \Omega_N$ a *word*, and if $N = l(w)$ we call the word *reduced*. A reduced word for w_0 is a *long word*. Now if $v \in \mathcal{B}_\lambda$ let us apply f_{Ω_1} to v as many times as we can. That is, let b_1 be the largest integer such that $f_{\Omega_1}^{b_1} v \neq 0$. Then let b_2 be the largest integer such that $f_{\Omega_2}^{b_2} f_{\Omega_1}^{b_1} v \neq 0$. Let $v' = f_{\Omega_N}^{b_N} \cdots f_{\Omega_1}^{b_1} v$. We summarize this situation symbolically as follows:

$$v \left[\begin{array}{ccc} b_1 & \cdots & b_N \\ \Omega_1 & \cdots & \Omega_N \end{array} \right] v'. \tag{2.7}$$

We refer to this as a *path*.

The crystal $\mathcal{B}_\lambda$ has a unique highest weight vector v_{high} such that $\text{wt}(v_{\text{high}}) = \lambda$, and a unique lowest weight vector v_{low} such that $\text{wt}(v_{\text{low}}) = w_0(\lambda)$. Note that $w_0(\lambda) = (\lambda_{r+1}, \cdots, \lambda_1)$.

LEMMA 2.1 *If $w = w_0$ then (2.7) implies that $v' = v_{\text{low}}$. In this case the integers $(b_1, \cdots, b_N)$ determine the vector v.*

Proof. See Littelmann [57] or Berenstein and Zelevinsky [4] for the proof that $v' = v_{\text{low}}$. (We are using f_i instead of the e_i that Littelmann uses, but the methods of proof are essentially unchanged.) Alternatively, the reader may prove this directly by pushing the arguments in Proposition 2.2 below a bit further. The fact that the b_i determine v follows from $v_{\text{low}} = f_{\Omega_N}^{b_N} \cdots f_{\Omega_1}^{b_1} v$ since then $v = e_{\Omega_1}^{b_1} \cdots e_{\Omega_N}^{b_N} v_{\text{low}}$. $\square$

The Gelfand-Tsetlin pattern can be recovered intrinsically from the location of a vector in the crystal as follows. Assume (2.7) with $w = w_0$. Let

$$\text{BZL}_\Omega(v) = \text{BZL}_\Omega^{(f)}(v) = (b_1, b_2, \cdots, b_N). \tag{2.8}$$

There are many reduced words representing w_0, but two will be of particular concern for us. If either

$$\Omega = \Omega_\Gamma = (1, 2, 1, 3, 2, 1, \cdots, r, r-1, \cdots, 3, 2, 1) \qquad (2.9)$$

or

$$\Omega = \Omega_\Delta = (r, r-1, r, r-2, r-1, r, \cdots, 1, 2, 3, \cdots, r), \qquad (2.10)$$

then Littelmann showed that

$$b_1 \geqslant 0 \qquad (2.11)$$
$$b_2 \geqslant b_3 \geqslant 0$$
$$b_4 \geqslant b_5 \geqslant b_6 \geqslant 0$$
$$\vdots$$

and that these inequalities characterize the possible patterns BZL_Ω. See in particular Theorem 4.2 of Littelmann [57], and Theorem 5.1 for this exact statement for Ω_Γ.

PROPOSITION 2.2 *Let $v \longmapsto \mathfrak{T} = \mathfrak{T}_v$ be the bijection defined by (2.6) from $\mathcal{B}_\lambda$ to the set of Gelfand-Tsetlin patterns with top row λ. (i) If Ω is the word Ω_Γ defined by (2.9) in Lemma 2.1, so*

$$v \begin{bmatrix} b_1 & b_2 & b_3 & b_4 & b_5 & b_6 & \cdots & b_{N-r+1} & b_{N-r+2} & \cdots & b_{N-1} & b_N \\ 1 & 2 & 1 & 3 & 2 & 1 & \cdots & r & r-1 & \cdots & 2 & 1 \end{bmatrix} v_{\text{low}},$$

where $N = \frac{1}{2}r(r+1)$. Then, with $\Gamma(\mathfrak{T})$ as defined in (1.11),

$$\Gamma(\mathfrak{T}_v) = \begin{bmatrix} \ddots & \vdots & \vdots & \vdots \\ & b_4 & b_5 & b_6 \\ & & b_2 & b_3 \\ & & & b_1 \end{bmatrix}.$$

(ii) If Ω is the word Ω_Δ defined in (2.10) in Lemma 2.1, so

$$v \begin{bmatrix} z_1 & z_2 & z_3 & z_4 & z_5 & z_6 & \cdots & z_{N-r+1} & z_{N-r+2} & \cdots & z_{N-1} & z_N \\ r & r-1 & r & r-2 & r-1 & r & \cdots & 1 & 2 & \cdots & r-1 & r \end{bmatrix} v_{\text{low}},$$

then, with $\Delta(\mathfrak{T})$ as in (1.11),

$$\Delta(q_r \mathfrak{T}_v) = \begin{bmatrix} \vdots & \vdots & \vdots & \ddots \\ z_6 & z_5 & z_4 & \\ z_3 & z_2 & & \\ z_1 & & & \end{bmatrix}.$$

(iii) We have $\mathfrak{T}_{\text{Sch}(v)} = q_r \mathfrak{T}_v$.

Proof. Most of this is in Littelmann [57], Berenstein and Zelevinsky [4] and Kirillov and Berenstein [51]. However it is also possible to see this directly from Kashiwara's description of the root operators by translating to tableaux, and so we will explain this.

Let $\Omega = \Omega_\Gamma$. We consider a Gelfand-Tsetlin pattern

$$\mathfrak{T} = \mathfrak{T}_v = \left\{ \begin{array}{ccccc} \ddots & & \vdots & & \iddots \\ & a_{r-1,r-1} & & a_{r-1,r} & \\ & & a_{rr} & & \end{array} \right\}$$

with corresponding tableau $\mathcal{T}$. Then a_{rr} is the number of 1's in $\mathcal{T}$, all of which must occur in the first row since $\mathcal{T}$ is column strict. In the tensor these correspond to $\boxed{1}$'s. Applying f_1 will turn some of these to $\boxed{2}$'s. In fact it follows from the definitions that the number b_1 of times that f_1 can be applied is the number of 1's in the first row of $\mathcal{T}$ that are not above a 2 in the second row. Now the number of 2's in the second row is $a_{r-1,r}$. Thus $b_1 = a_{rr} - a_{r-1,r}$.

For example if

$$\mathfrak{T} = \left\{ \begin{array}{ccccccc} 10 & & 5 & & 3 & & 0 \\ & 9 & & 4 & & 2 & \\ & & 7 & & 3 & & \\ & & & 5 & & & \end{array} \right\} \quad \leftrightarrow \quad \mathcal{T} = \begin{array}{|c|c|c|c|c|c|c|c|c|c|} \hline 1 & 1 & 1 & 1 & 1 & 2 & 2 & 3 & 3 & 4 \\ \hline \end{array}$$

then we can apply f_1 twice (so $b_1 = a_{33} - a_{23} = 5 - 3 = 2$) and we obtain

$$\begin{array}{|c|c|c|c|c|c|c|c|c|c|} \hline 1 & 1 & 1 & 2 & 2 & 2 & 2 & 3 & 3 & 4 \\ \hline \end{array}$$

Now b_2 is the number of times we can apply f_2. This will promote $\boxed{2} \longrightarrow \boxed{3}$ but only if the 2 in the tableau is not directly above a 3. One 2 will be promoted from the second row ($1 = a_{23} - a_{13} = 3 - 2$) and three will be promoted from the first row ($3 = a_{22} - a_{12} = 7 - 4$). Thus $b_2 = a_{22} + a_{23} - a_{12} - a_{13}$. This produces the tableau

$$\begin{array}{|c|c|c|c|c|c|c|c|c|c|} \hline 1 & 1 & 1 & 2 & 3 & 3 & 3 & 3 & 3 & 4 \\ \hline \end{array}$$

After this, we can apply f_1 once ($1 = a_{23} - a_{13} = 3 - 2$) promoting one 1 and giving

$$\begin{array}{|c|c|c|c|c|c|c|c|c|c|} \hline 1 & 1 & 2 & 2 & 3 & 3 & 3 & 3 & 3 & 4 \\ \hline \end{array}$$

Thus $b_3 = a_{23} - a_{13}$. After this, we apply f_3 seven times promoting two 3's in the third row ($2 = 2 - 0 = a_{31} - a_{30}$), one 3 in the second row ($1 = 4 - 3 = a_{12} - a_{02}$), and four 3's in the first row ($4 = 9 - 5 = a_{11} - a_{01}$) to obtain

$$\begin{array}{|c|c|c|c|c|c|c|c|c|c|} \hline 1 & 1 & 2 & 2 & 3 & 4 & 4 & 4 & 4 & 4 \\ \hline \end{array}$$

Thus $b_4 = a_{11} + a_{12} + a_{13} - a_{01} - a_{02} - a_{03}$. One continues in this way.

From this discussion (i) is clear. We refer to Kirillov and Berenstein [51] for the computation of the involution Sch. Thus we will refer to [51] for (iii) and using (iii) we will prove (ii). By (iii) and (2.5) the map $\phi_\lambda : \mathcal{B}_\lambda \longrightarrow \mathcal{B}_{-w_0\lambda}$ satisfying (2.3) has the effect

$$\mathfrak{T}_{\phi_\lambda v} = q_r \mathfrak{T}_{\psi_v} = (-q_r \mathfrak{T}_v)^{\mathrm{rev}}. \tag{2.12}$$

Since ϕ_λ changes f_i to f_{r+1-i} it replaces $b_1, \cdots, b_N$ (computed for $\phi_\lambda v \in \mathcal{B}_{-w_0\lambda}$) by $z_1, \cdots, z_N$. It is easy to see from the definition (1.11) that

$$\Gamma(-\mathfrak{T}^{\mathrm{rev}}) = \Delta(\mathfrak{T})^{\mathrm{rev}},$$

and (ii) follows. $\qquad\qquad\qquad\qquad\qquad\qquad\qquad\qquad\qquad\qquad\qquad\qquad\qquad\square$

Now let us reinterpret the factors $G_\Gamma(\mathfrak{T}_v)$ and $G_\Delta(q_r\mathfrak{T}_v)$ defined in Chapter 1. It follows from Proposition 2.2 that the numbers Γ_{ij} and Δ_{ij} that appear in the respective arrays are exactly the quantities that appear in $\mathrm{BZL}_\Omega(v)$ when $\Omega = \Omega_\Gamma$ or Ω_Δ, and we have only to describe the circling and boxing decorations.

The circling is clear: we circle b_i if either $i \in \{1, 3, 6, 10, \cdots\}$ (so b_i is the first element of its row) and $b_i = 0$, or if $i \notin \{1, 3, 6, 10, \cdots\}$ and $b_i = b_{i+1}$. This is a direct translation of the circling definition in Chapter 1.

Let us illustrate this with an example. In Figure 2.1 we compute $\Gamma(\mathfrak{T}_v)$ for a vertex of the A_2 crystal with highest weight $(5,3,0)$. We have

$$v \begin{bmatrix} 0 & & 2 & & 2 \\ & 1 & & 2 & & 1 \end{bmatrix} v_{\mathrm{low}}$$

so that $b_1 = 0$, $b_2 = b_3 = 2$. The inequalities (2.11) assert that $b_1 \geqslant 0$ and $b_2 \geqslant b_3 \geqslant 0$. Since two of these are sharp, we circle b_1 and b_2 and

$$\Gamma(\mathfrak{T}_v) = \begin{bmatrix} ② & 2 \\ & ⓪ \end{bmatrix}, \qquad \mathfrak{T}_v = \left\{ \begin{array}{ccccc} 5 & & 3 & & 0 \\ & 3 & & 2 & \\ & & 2 & & \end{array} \right\}.$$

As for the boxing, the condition has an interesting reformulation in terms of the crystal, which we describe next. Consider the path

$$v, f_{\Omega_1} v, f_{\Omega_1}^2 v, \cdots, f_{\Omega_1}^{b_1} v, f_{\Omega_2} f_{\Omega_1}^{b_1} v, \cdots, f_{\Omega_2}^{b_2} f_{\Omega_1}^{b_1} v, \cdots, f_{\Omega_N}^{b_N} \cdots f_{\Omega_1}^{b_1} v = v_{\mathrm{low}}$$

through the crystal from v to v_{low}. The b_j are the lengths of consecutive moves along edges f_{Ω_j} in the path. (These are depicted by straight-line segments in the figure.) If u is any vertex and $1 \leqslant i \leqslant r$, then the i-string through u is the set of vertices that can be obtained from u by repeatedly applying either e_i or f_i. The boxing condition then amounts to the assumption that the canonical path contains the entire Ω_t string through $f_{\Omega_{t-1}}^{b_{t-1}} \cdots f_{\Omega_1}^{b_1} v$. That is, the condition for b_t to be boxed is that

$$e_{\Omega_t} f_{\Omega_{t-1}}^{b_{t-1}} \cdots f_{\Omega_1}^{b_1} v = 0. \tag{2.13}$$

Consider the following example. Let $\lambda = (4, 2, 0)$, and let $\Omega = \Omega_\Gamma = (1, 2, 1)$. Then, as pictured in Figure 2.2, we have $b_1 = 2$, $b_2 = 3$ and $b_3 = 1$. Since the path

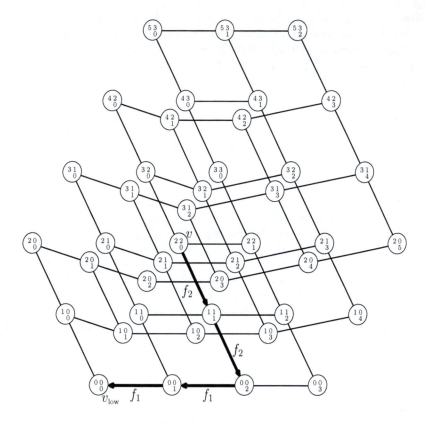

Figure 2.1 **The circling rule.** The crystal graph has highest weight $\lambda = (5,3,0)$. The element v_{low} has lowest weight $w_0\lambda = (0,3,5)$, and v has weight $(2,0,3)$. The labels of the vertices are the Γ arrays. The word $\Omega_\Gamma = 121$ is used to compute $\Gamma(\mathfrak{T}_v)$. The root operator f_1 moves left along crystal edges, and the root operator f_2 moves down and to the right. The crystal has been drawn so that elements of a given weight are placed in diagonally aligned clusters.

includes the entire 2-string through $f_1^2 v$ (or equivalently, since $e_2 f_1^2 v = 0$) we box b_2 and

$$\Gamma(\mathfrak{T}_v) = \left\{ \begin{array}{cc} \boxed{3} & 1 \\ & 2 \end{array} \right\}, \qquad \mathfrak{T}_v = \left\{ \begin{array}{ccc} 4 & 2 & 0 \\ & 4 & 1 \\ & & 3 \end{array} \right\}.$$

It is not hard to see that the decorations of $\Gamma(\mathfrak{T}_v)$ described this way agree with those already defined in Chapter 1.

Now that we have explained the boxing and circling rules geometrically for the

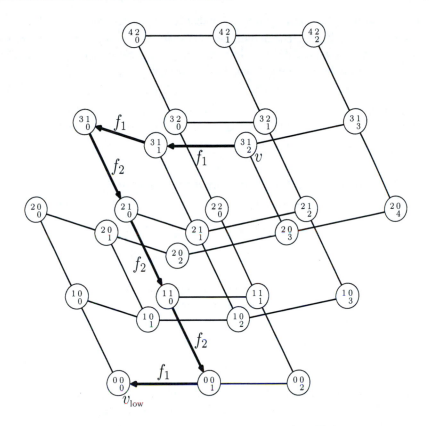

Figure 2.2 **The boxing rule.** The crystal graph $\mathcal{B}_\lambda$ with $\lambda = (4, 2, 0)$. The element v_{low} has lowest weight $w_0\lambda = (0, 2, 4)$, and v has weight $(3, 2, 1)$. The word $\Omega_\Gamma = 121$ is used to compute $\Gamma(\mathfrak{T}_v)$. The root operator f_1 moves left along edges, and the root operator f_2 moves down and to the right. The crystal has been drawn so that elements of a given weight are placed in diagonally aligned clusters.

BZL pattern, it is natural to make the definition

$$G_\Gamma(v) = G_\Gamma^{(f)}(v) = \prod_{i=1}^{\frac{1}{2}r(r+1)} \begin{cases} q^{b_i} & \text{if } b_i \text{ is circled,} \\ g(b_i) & \text{if } b_i \text{ is boxed,} \\ h(b_i) & \text{if } b_i \text{ is neither circled nor boxed,} \\ 0 & \text{if } b_i \text{ is both circled and boxed,} \end{cases}$$

(2.14)

with a reduced version

$$G_\Gamma^\flat(v) = G_\Gamma^{\flat(f)}(v) = \prod_{i=1}^{\frac{1}{2}r(r+1)} \begin{cases} 1 & \text{if } b_i \text{ is circled,} \\ g^\flat(b_i) & \text{if } b_i \text{ is boxed,} \\ h^\flat(b_i) & \text{if } b_i \text{ is neither circled nor boxed,} \\ 0 & \text{if } b_i \text{ is both circled and boxed.} \end{cases}$$

(2.15)

Here the b_i are as in Proposition 2.2 (i) and the functions $g^\flat$ and $h^\flat$ are as in (1.23).

The (f) in this notation is optionally added to emphasize that we are basing this description on $\text{BZL}^{(f)}$. The definitions of $G_\Delta(v) = G_\Delta^{(f)}(v)$ and $G_\Delta^\flat(v)$ are identical, except that we use the z_i in Proposition 2.2 (ii). Now by Proposition 2.2 and the above discussion of boxing and circling, we have

$$G_\Gamma(v) = G_\Gamma(\mathfrak{T}_v), \qquad G_\Delta(v) = G_\Delta(\mathfrak{T}_v). \tag{2.16}$$

In order to finish describing the p-part of the multiple Dirichlet series in terms of crystal graphs we need to describe k_Γ and k_Δ. If $\mathfrak{T}$ is a Gelfand-Tsetlin pattern, let $d_i = d_i(\mathfrak{T})$ be the sum of the i-th row of $\mathfrak{T}$, so that if $\mathfrak{T}$ is as in (1.5), then $d_i = \sum_{j=i}^r a_{ij}$.

PROPOSITION 2.3 *If $v \in \mathcal{B}_\lambda$ and d_i are the row sums of $\mathfrak{T}_v$, then*

$$\text{wt}(v) = (d_r, d_{r-1} - d_r, \cdots, d_0 - d_1). \tag{2.17}$$

Moreover if $b_1, \cdots, b_{\frac{1}{2}r(r+1)}$ are as in Proposition 2.2 (i), then

$$\langle \text{wt}(v) - w_0(\lambda), \rho \rangle = \sum_i b_i. \tag{2.18}$$

Proof. Let $b_1, \cdots, b_{\frac{1}{2}r(r+1)}$ be as in Proposition 2.2 (i). Thus they are the entries in $\Gamma(\mathfrak{T})$. Since v_{low} is obtained from v by applying f_1 $b_1 + b_3 + b_6 + \ldots$ times, f_2 $b_2 + b_5 + b_9$ times, etc., we have

$$\text{wt}(v) - w_0(\lambda) = \text{wt}(v) - \text{wt}(v_{\text{low}}) = k_r \alpha_1 + k_{r-1}\alpha_2 + \ldots + k_1 \alpha_r \tag{2.19}$$

where $k_r = b_1 + b_3 + b_6 + \ldots$, $k_{r-1} = b_2 + b_5 + \ldots$. Now (2.18) follows since $\langle \alpha_i, \rho \rangle = 1$.

By Proposition 2.2 (i) the b_i are the entries in $\Gamma(\mathfrak{T}_v)$. In particular $b_1 = a_{r,r} - a_{r-1,r}$, $b_3 = a_{r-1,r} - a_{r-2,r}$, etc. and so the sum defining k_r collapses. Since $a_{0,i} = \lambda_{i+1}$ we have

$$k_r = a_{r,r} - a_{0,r} = d_r - \lambda_{r+1}.$$

Similarly

$$k_{r-1} = d_{r-1} - \lambda_r - \lambda_{r+1}$$
$$k_{r-2} = d_{r-2} - \lambda_{r-1} - \lambda_r - \lambda_{r+1}$$
$$\vdots$$

In view of (1.8) and the definitions of the previous chapter, $k_i = k_{\Gamma,i}$.

Remembering that $\text{wt}(v_{\text{low}}) = w_0(\lambda) = (\lambda_{r+1}, \lambda_r, \cdots, \lambda_1)$, this shows that

$$\text{wt}(v) - (\lambda_{r+1}, \lambda_r, \cdots, \lambda_1) =$$
$$(d_r - \lambda_{r+1})\alpha_1 + (d_{r-1} - \lambda_r - \lambda_{r+1})\alpha_2 + \ldots + (d_1 - \lambda_2 - \ldots - \lambda_r)\alpha_r =$$
$$(d_r - \lambda_{r+1}, d_{r-1} - d_r - \lambda_r, \cdots, -d_1 + \lambda_2 + \ldots + \lambda_{r+1}).$$

Since $d_0 = \lambda_1 + \ldots + \lambda_{r+1}$, we get (2.17). $\qquad\square$

PROPOSITION 2.4 *Let $v \in \mathcal{B}_\lambda$, and let $k_i = k_{\Gamma,i}(\mathfrak{T}_v)$. Then the k_i are the unique integers such that*

$$\sum_{i=1}^r k_i \alpha_i = \lambda - w_0(\text{wt}(v)). \tag{2.20}$$

Proof. Applying w_0 to (2.19) gives (2.20) since $-w_0\alpha_i = \alpha_{r+1-i}$. The uniqueness of the k_i comes from the linear independence of the k_i. $\qquad\square$

THEOREM 2.5 *We have*

$$H_\Gamma(p^{k_1}, \cdots, p^{k_r}; p^{l_1}, \cdots, p^{l_r}) = \sum_{\substack{\in \mathcal{B}_{\lambda+\rho} \\ \mathrm{wt}(v)=w_0\left(\lambda-\sum_i k_i\alpha_i\right)}} G_\Gamma(v),$$

$$H_\Delta(p^{k_1}, \cdots, p^{k_r}; p^{l_1}, \cdots, p^{l_r}) = \sum_{\substack{v \in \mathcal{B}_{\lambda+\rho} \\ \mathrm{wt}(v)=w_0\left(\lambda-\sum_i k_i\alpha_i\right)}} G_\Delta(v),$$

where $\lambda = (\lambda_1, \cdots, \lambda_{r+1})$ *with* $\lambda_i = \sum_{j \geqslant i} l_i.$

In view of Theorem 1.2, these two expressions are equal.

Proof. This is clear in view of (2.16) and (2.20). $\qquad\square$

There are also reduced versions of these equalities:

$$H_\Gamma^\flat(p^{k_1}, \cdots, p^{k_r}; p^{l_1}, \cdots, p^{l_r}) = \sum_{\substack{v \in \mathcal{B}_{\lambda+\rho} \\ \mathrm{wt}(v)=w_0\left(\lambda-\sum_i k_i\alpha_i\right)}} G_\Gamma^\flat(v),$$

$$H_\Delta^\flat(p^{k_1}, \cdots, p^{k_r}; p^{l_1}, \cdots, p^{l_r}) = \sum_{\substack{v \in \mathcal{B}_{\lambda+\rho} \\ \mathrm{wt}(v)=w_0\left(\lambda-\sum_i k_i\alpha_i\right)}} G_\Delta^\flat(v).$$

Chapter Three

Duality

The material contained in this chapter won't be used again until Chapter 18 and so can be skipped without loss of continuity. Its purpose is to point out that the boxing and circling decorations of the BZL patterns that were introduced in the last chapter are in a sense dual to each other.

In Chapter 2 we used either of the notations

$$v \begin{bmatrix} b_1 & \cdots & b_N \\ i_1 & \cdots & i_N \end{bmatrix} v' \qquad \text{or} \qquad v \begin{bmatrix} b_1 & \cdots & b_N \\ i_1 & \cdots & i_N \end{bmatrix}^{(f)} v'$$

to mean that $v' = f_{i_N}^{b_N} \cdots f_{i_1}^{b_1} v$ where each integer b_k is as large as possible in the sense that $f_{i_k}^{b_k+1} f_{i_{k-1}}^{b_{k-1}} \cdots f_{i_1}^{b_1} v = 0$. In this chapter, we will exclusively use the second notation – the superscript (f) will be needed to avoid confusion since we now analogously define

$$v \begin{bmatrix} b_1 & \cdots & b_N \\ i_1 & \cdots & i_N \end{bmatrix}^{(e)} v' \tag{3.1}$$

to mean that $v' = e_{i_N}^{b_N} \cdots e_{i_1}^{b_1} v$ where $e_{i_k}^{b_k+1} e_{i_{k-1}}^{b_{k-1}} \cdots e_{i_1}^{b_1} v = 0$ for all $1 \leqslant k \leqslant N$. Thus

$$v \begin{bmatrix} b_1 & \cdots & b_N \\ i_1 & \cdots & i_N \end{bmatrix}^{(e)} v' \qquad \text{if and only if} \qquad v' \begin{bmatrix} b_N & \cdots & b_1 \\ i_N & \cdots & i_1 \end{bmatrix}^{(f)} v.$$

Let us assume that (3.1) holds, where $\Omega = (i_1, \cdots, i_N)$ and $N = \frac{1}{2}r(r+1)$. Assuming that $\Omega = \Omega_\Gamma$ or Ω_Δ, defined in (2.9) and (2.10), we may decorate the values b_k according to circling and boxing rules that are analogous to those defined in Chapter 2. Thus b_k is circled if and only if either $b_k = 0$ (when k is a triangular number) or $b_k = b_{k+1}$ (when it is not). And b_k is boxed if and only if $f_{i_k} e_{i_{k-1}}^{b_{k-1}} \cdots e_{i_1}^{b_1} v = 0$, which is equivalent to saying that the path (3.1) includes the entire i_k string through $e_{i_{k-1}}^{b_{k-1}} \cdots e_{i_1}^{b_1} v$. In this case, dual to Lemma 2.1, we have $v' = v_{\text{high}}$.

The definition of the multiple Dirichlet series can be made equally well with respect to the e_i. Indeed, we adapt (2.14) and define

$$G_\Delta^{(e)}(v) = \prod_{i=1}^{\frac{1}{2}r(r+1)} \begin{cases} q^{b_i} & \text{if } b_i \text{ is circled,} \\ g(b_i) & \text{if } b_i \text{ is boxed,} \\ h(b_i) & \text{if } b_i \text{ is neither circled nor boxed,} \\ 0 & \text{if } b_i \text{ is both circled and boxed,} \end{cases}$$

assuming that $\Omega = \Omega_\Delta$; if instead $\Omega = \Omega_\Gamma$, then $G_\Gamma^{(e)}$ is defined by the same formula.

PROPOSITION 3.1

$$H_\Gamma(p^{k_1}, \cdots, p^{k_r}; p^{l_1}, \cdots, p^{l_r}) = \sum_{\substack{v \in \mathcal{B}_{\lambda+\rho} \\ \mathrm{wt}(v) = \lambda - \sum_i k_i \alpha_i}} G_\Delta^{(e)}(v).$$

$$H_\Delta(p^{k_1}, \cdots, p^{k_r}; p^{l_1}, \cdots, p^{l_r}) = \sum_{\substack{v \in \mathcal{B}_{\lambda+\rho} \\ \mathrm{wt}(v) = \lambda - \sum_i k_i \alpha_i}} G_\Gamma^{(e)}(v).$$

Proof. We replace v by $\mathrm{Sch}(v)$ in Theorem 2.5. It follows from (2.1) that

$$G_\Delta^{(e)}(\mathrm{Sch}(v)) = G_\Gamma^{(f)}(v),$$

and $\mathrm{wt}(\mathrm{Sch}(v)) = w_0(\mathrm{wt}(v))$, and the statement follows.   □

The circling and boxing rules seem quite different from each other, but actually they are closely related, and the involution sheds light on this fact also. Let us apply Proposition 2.2 (i) to $\mathfrak{T}_v$ and Proposition 2.2 (ii) to $\mathfrak{T}_v = q_r \mathfrak{T}_{\mathrm{Sch}(v)}$. For the latter, we see that

$$\Delta(\mathfrak{T}_v) = \begin{bmatrix} \vdots & \vdots & \vdots & \ddots \\ l_6 & l_5 & l_4 & \\ l_3 & l_2 & & \\ l_1 & & & \end{bmatrix},$$

where by (2.1)

$$\mathrm{Sch}(v) \begin{bmatrix} l_1 & l_2 & l_3 & l_4 & l_5 & l_6 & \cdots & l_{N-2} & l_{N-1} & l_N \\ r & r-1 & r & r-2 & r-1 & r & \cdots & r-2 & r-1 & r \end{bmatrix}^{(f)}_{v_{\mathrm{low}}}.$$

Now by definition an entry is circled in $\Gamma(\mathfrak{T}_v)$ if and only if the corresponding entry is boxed in $\Delta(\mathfrak{T}_v)$. If we use the word Ω_Γ for v and Ω_Δ for $\mathrm{Sch}(v)$, then we obtain two BZL patterns in which circled entries in one correspond to boxed entries in the other!

Figure 3.1 illustrates this with an example. The two equalities marked in $\mathfrak{T}_v$ give rise to two circles in $\Gamma(\mathfrak{T}_v)$ and two boxes in $\Delta(\mathfrak{T}_v)$. We can see these in the marked paths from v and $\mathrm{Sch}(v)$ to v_{low}.

But there is another way to look at this. It follows from (2.1) that

$$v \begin{bmatrix} l_1 & l_2 & l_3 & l_4 & l_5 & l_6 & \cdots & l_{N-2} & l_{N-1} & l_N \\ 1 & 2 & 1 & 3 & 2 & 1 & \cdots & 3 & 2 & 1 \end{bmatrix}^{(e)}_{v_{\mathrm{high}}}.$$

This means that we may generate $\Gamma(\mathfrak{T}_v)$ and $\Delta(\mathfrak{T}_v)$ from the same element v and the same word Ω_Γ, but applying f_i successively to generate the entries b_i of $\Gamma(\mathfrak{T}_v)$ and applying the e_i (in the same order) to generate the entries l_i of $\Delta(\mathfrak{T}_v)$. The boxing and circling rules are defined analogously for the $\Delta(\mathfrak{T}_v)$ as for $\Gamma(\mathfrak{T}_v)$. There exists a natural *box-circle duality*: a bijection between the b_i and the l_i in which b_i is circled if and only if the corresponding l_i is boxed. Note that this bijection changes the order:

$$\begin{array}{cccccccc} b_1 & b_2 & b_3 & b_4 & b_5 & b_6 & \cdots \\ \updownarrow & \updownarrow & \updownarrow & \updownarrow & \updownarrow & \updownarrow & \\ l_1 & l_3 & l_2 & l_6 & l_5 & l_4 & \cdots \end{array}$$

This rather striking property of the crystal graph is illustrated in Figures 3.1 and 3.2.

$$\mathfrak{T}_v = \left\{ \begin{array}{ccc} 5 & 3 & 0 \\ & 3 \quad\ \ 2 & \\ & 2 & \end{array} \right\}$$

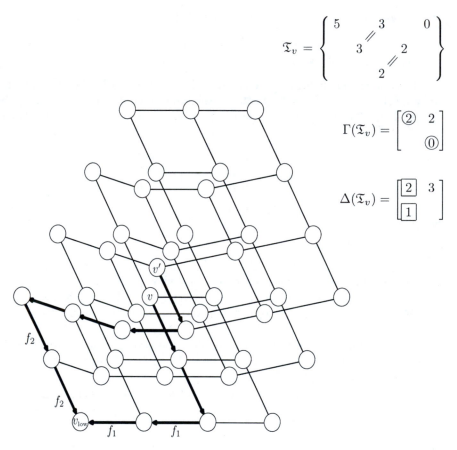

$$\Gamma(\mathfrak{T}_v) = \begin{bmatrix} ② & 2 \\ & ⓪ \end{bmatrix}$$

$$\Delta(\mathfrak{T}_v) = \begin{bmatrix} \boxed{2} & 3 \\ \boxed{1} & \end{bmatrix}$$

Figure 3.1 **Box-Circle Duality.** Here v is the marked element of $\mathcal{B}_\lambda$, $\lambda = (5, 3, 0)$, and $v' = \mathrm{Sch}(v)$. With $\mathfrak{T} = \mathfrak{T}_v$, $\Gamma(\mathfrak{T})$ and $\Delta(\mathfrak{T})$ are obtained from v and v' as BZL patterns for the words $\Omega_\Gamma = (1, 2, 1)$ and $\Omega_\Delta = (2, 1, 2)$. Boxes in one array correspond to circles in the other.

$$\mathfrak{T}_v \;=\; \left\{ \begin{array}{ccc} 5 & 3 & 0 \\ & 3 \quad 2 & \\ & 2 & \end{array} \right\}$$

$$\Gamma(\mathfrak{T}_v) = \begin{bmatrix} ② & 2 \\ & ⓪ \end{bmatrix}$$

$$\Delta(\mathfrak{T}_v) = \begin{bmatrix} \boxed{2} & 3 \\ \boxed{1} & \end{bmatrix}$$

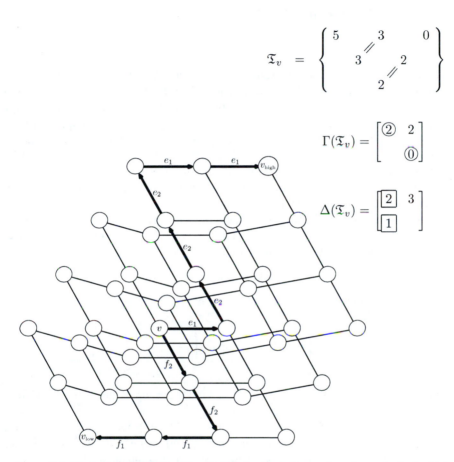

Figure 3.2 **More Box-Circle Duality.** Continuing from the previous figure, instead of taking the f-path from $v' = \mathrm{Sch}(v)$ we may take the e-path from v and obtain the same decorated BZL string. As before, With $\mathfrak{T} = \mathfrak{T}_v$, $\Gamma(\mathfrak{T})$ and $\Delta(\mathfrak{T})$ are obtained from v as BZL patterns for the word $\Omega_\Gamma = (1, 2, 1)$, but using f root operators for Γ and e root operators for Δ. The circling of entries in one path corresponds to boxing of entries in the other path, a striking combinometrical property of the crystal.

Chapter Four

Whittaker Functions

Weyl group multiple Dirichlet series are expected to be Whittaker coefficients of metaplectic Eisenstein series. The fact that Whittaker coefficients of Eisenstein series reduce to the crystal description that we gave in Chapter 2 is proved for Type A. In a classical setting, this was established by Brubaker, Bump, and Friedberg [13]. Alternatively, on the adele group, the corresponding local computation reduces to the evaluation of a type of p-adic integral. These were considered by McNamara [65], who reduced the integrals to sums over crystals by a very interesting method. A full treatment of this topic is outside the scope of this work, but we will introduce this subject by considering the case where $n = 1$.

In this chapter only, we will use F to denote a nonarchimedean local field and $\boldsymbol{F}$ a global field. Let $\hat{G}(\mathbb{C}) = \mathrm{GL}_{r+1}(\mathbb{C})$, which is the L-group of $G = \mathrm{GL}_{r+1}$. Let T be the diagonal torus in G, and let $\hat{T}(\mathbb{C})$ be the diagonal torus in $\hat{G}(\mathbb{C})$. If

$$z = \mathrm{diag}(z_1, \cdots, z_{r+1}) \in \hat{T}(\mathbb{C}),$$

then let χ_z be the unramified quasicharacter of $T(F)$ defined by

$$\chi_z(t) = \prod z_i^{\mathrm{ord}(t_i)}, \qquad t = \begin{pmatrix} t_1 & & \\ & \ddots & \\ & & t_{r+1} \end{pmatrix}.$$

Then $z \longmapsto \chi_z$ is an isomorphism of $\hat{T}(\mathbb{C})$ with the group of unramified quasicharacters of $T(F)$.

We will denote by W the Weyl group of G or $\hat{G}$. Each Weyl group element has a representative w that is a permutation matrix, and for definiteness if $w \in W$ we will use that representative. Let w_0 be the long Weyl group element.

Let $B(F)$ be the Borel subgroup of upper triangular matrices, and let $U(F)$ be the subgroup of upper triangular unipotent matrices, so $B(F) = T(F)U(F)$. Given a quasicharacter χ of $T(F)$, we may extend it to a character of $B(F)$ by letting $U(F)$ be in the kernel. Let $\delta : T(F) \longrightarrow \mathbb{R}^\times$ be the quasicharacter $\delta(t) = \prod_{i=1}^{r+1} |t_i|^{r+2-2i}$. Extended to $B(F)$, this is the modular character.

Let V_χ be the space of locally constant functions f on $G(F)$ such that

$$f(bg) = \delta^{1/2} \chi(b) f(g), \qquad b \in B(F).$$

Then $G(F)$ acts by right translation: that is, $\pi_\chi : G(F) \longrightarrow \mathrm{End}(V_\chi)$ is defined by $\pi_\chi(g)f(x) = f(xg)$. If $\chi = \chi_z$ we will also write (π_z, V_z) for the representation (π_χ, V_χ). The representations π_z are called the *unramified principal series*. They are irreducible if z is in general position.

Let $K = \mathrm{GL}_{r+1}(\mathfrak{o})$, where $\mathfrak{o}$ is the ring of integers in F. Then $G(F) = B(F)\,K$. Let $f_{\mathbf{z}}^{\circ}$ be the *normalized spherical vector* in $V_{\mathbf{z}}$, defined by

$$f_{\mathbf{z}}^{\circ}(bk) = \delta^{1/2}\chi(b), \qquad b \in B(F),\ k \in K.$$

This is a K-fixed vector. It is known that any irreducible representation having a K-fixed vector is a subquotient of some $\pi_{\mathbf{z}}$, with $\mathbf{z}$ unique up to the action of the Weyl group. Therefore the quotient $\hat{T}(\mathbb{C})/W$ of $\hat{T}(\mathbb{C})$ by the action of W through conjugation parametrizes the spherical representations.

We recall that we are identifying $\mathbb{Z}^{r+1}$ with the weight lattice Λ of $\hat{G}(\mathbb{C})$. This is the group of *weights*, which are the rational characters of $\hat{T}(\mathbb{C})$. If $\mathbf{z} \in \hat{T}(\mathbb{C})$ and $\mu = (\mu_1, \cdots, \mu_{r+1}) \in \mathbb{Z}^{r+1}$ then with this identification $\mathbf{z}^{\mu} = \prod z_i^{\mu_i}$. Recall that $\lambda \in \Lambda$ is *dominant* if $\lambda_1 \geqslant \cdots \geqslant \lambda_{r+1}$. If λ is dominant then there is an irreducible module of $\hat{G}(\mathbb{C})$ with highest weight λ. If $\mathbf{z} \in \hat{T}(\mathbb{C})$ then the trace of $\mathbf{z}$ on this module is the Schur polynomial (Macdonald [64]) $s_{\lambda}(\mathbf{z}) = s_{\lambda}(z_1, \cdots, z_{r+1})$.

Let ψ_F be an additive character of F. Assume that the largest fractional ideal on which ψ_F is trivial is the ring $\mathfrak{o}$ of integers. Define $\psi : U(F) \longrightarrow \mathbb{C}$ by

$$\psi \begin{pmatrix} 1 & u_{12} & \cdots & u_{1,r+1} \\ & 1 & \cdots & u_{2,r+1} \\ & & \ddots & \vdots \\ & & & 1 \end{pmatrix} = \psi_F\left(\sum_{i=1}^{r} u_{i,i+1}\right). \qquad (4.1)$$

If $\lambda \in \Lambda = \mathbb{Z}^{r+1}$, let

$$t_{\lambda} = \begin{pmatrix} \varpi^{\lambda_1} & & \\ & \ddots & \\ & & \varpi^{\lambda_{r+1}} \end{pmatrix},$$

where ϖ is a prime element of $\mathfrak{o}$.

THEOREM 4.1 *If $|z_i/z_{i+1}| < 1$ for $i = 1, \cdots, r$ then*

$$\int_{U(F)} f_{\mathbf{z}}^{\circ}(w_0 u t_{\lambda})\psi(u)\,du =$$

$$\begin{cases} \left[\prod_{\alpha \in \Phi^+}(1 - q^{-1}\mathbf{z}^{\alpha})\right]\delta^{1/2}(t_{\lambda})s_{\lambda}(\mathbf{z}) & \text{if } \lambda \text{ is dominant,} \\ 0 & \text{otherwise.} \end{cases} \qquad (4.2)$$

This is the *Casselman-Shalika formula*. It was conjectured in a letter of Langlands to Godement. Except for the constant $\prod(1 - q^{-1}\mathbf{z}^{\alpha})$ it was proved first by Shintani [67]. The proof of Casselman and Shalika determines the constant, and contains important new ideas.

The assumption that $|z_i/z_{i+1}| < 1$ makes the integral absolutely convergent. For general $\mathbf{z}$ the integral may still be interpreted by analytic continuation or other renormalization procedure, and the formula is still correct.

Proof. See Casselman and Shalika [23]. □

For future reference, we mention an important variant. If we omit the additive character ψ from the integration, then we obtain the important *formula of Gindikin and Karpelevich*.

THEOREM 4.2 (Formula of Gindikin and Karpelevich) *If $|z_i/z_{i+1}| < 1$ for $i = 1, \cdots, r$ then*

$$\int_{U(F)} f^\circ(w_0 u t_\lambda)\, du = \left[\prod_{\alpha \in \Phi^+} \frac{1 - q^{-1}z^\alpha}{1 - z^\alpha}\right] (\delta^{1/2}\chi_{w_0 z})(t_\lambda).$$

This formula was first proved by Langlands [54] after a similar formula was proved in the archimedean case by Gindikin and Karpelevich. This formula computes the normalizing constant for the intertwining integral between two induced representations. It has many applications. It computes the constant terms of Eisenstein series, which control their poles. As such, it is basic in both the Langlands-Shahidi method and the Rankin-Selberg method. It also computes the Plancherel measure for the K-biinvariant part of $L^2(G(F))$ (see Macdonald [63]).

Proof. See Casselman [22], Theorem 3.1.   □

A generalization of the formula of Gindikin and Karpelevich to metaplectic covers of GL_{r+1} was found by Kazhdan and Patterson [49]. If $\tilde{G}(F)$ is the n-fold metaplectic cover of $\mathrm{GL}_{r+1}(F)$ then the cover splits over $U(F)$ and so we may embed $U(F) \longrightarrow \tilde{G}(F)$. In [49], Proposition I.2.4 they proved

$$\int_{U(F)} f^\circ(w_0 u)\, du = \prod_{\alpha \in \Phi^+} \frac{1 - q^{-1}z^{n\alpha}}{1 - z^{n\alpha}}. \tag{4.3}$$

We will reinterpret the formula of Gindikin and Karpelevich and its extension by Kazhdan and Patterson in terms of crystals in the last chapter of the book.

Returning to (4.2),

$$W(g) = \int_{U(F)} f^\circ_z(w_0 u g)\psi(u)\, du$$

is called the *spherical Whittaker function*. We see that the values of the Whittaker function are Schur polynomials multiplied by the normalization constant $\prod(1 - q^{-1}z^\alpha)$.

Now let F be a global field and $\mathbb{A}$ its adele ring. Let $\zeta = (\zeta_1, \cdots, \zeta_{r+1})$ be complex numbers. Define a quasicharacter of $\chi_\zeta : B(\mathbb{A}) \longrightarrow \mathbb{C}^\times$ by

$$\chi_\zeta \begin{pmatrix} t_1 & * & \cdots & * \\ & t_2 & & * \\ & & \ddots & \vdots \\ & & & t_{r+1} \end{pmatrix} = \prod_{i=1}^{r+1} |t_i|^{\zeta_i}.$$

Let $V = V_\zeta$ be the space of smooth functions f on $G(\mathbb{A}) = \mathrm{GL}_{r+1}(\mathbb{A})$ that satisfy

$$f(bg) = \delta^{1/2}\chi_\zeta(b)f(g),$$

with the group acting in a representation π by right translation: $(\pi(g)f)(h) = f(hg)$. We may decompose this representation as a restricted tensor product $\pi = \otimes_v \pi_v$ on a space $\otimes_v V_v$ over the places v of F. We recall what this means. If v is nonarchimedean then V_v is what was previously denoted V_z where $z = z(v, \zeta) =$

$(z_1, \cdots, z_{r+1})$ with $z_i = q_v^{-\zeta_v}$, and q_v is the residue field cardinality. Every element of V is a finite linear combination of functions of the form $\otimes_v f_v$ with $f_v \in V_v$. This is the function

$$f(g) = \prod_v f_v(g_v)$$

for adeles $g = (g_v)$. The vector $f_v = f^\circ_{z(v,\zeta)}$ and $g_v \in GL_{r+1}(\mathfrak{o}_v)$ for almost all v, so all but finitely many factors in the product equal 1. If $f = \otimes_v f_v$ then we say that f is a *pure tensor*.

If $f_\zeta \in V_\zeta$ then we may consider the Eisenstein series

$$E_\zeta(g, f_\zeta) = \sum_{\gamma \in B(F)\backslash G(F)} f_\zeta(\gamma g).$$

The series is convergent if $\zeta_i - \zeta_{i+1} > 1$ for $1 \leqslant i \leqslant r$. It has meromorphic continuation by the theory of Selberg and Langlands.

Now let $\psi_\mathbb{A} : \mathbb{A} \longrightarrow \mathbb{C}$ be a character that is trivial on F. If v is a place of F let $\psi_v : F_v \longrightarrow \mathbb{C}$ be the restriction of $\psi_\mathbb{A}$ to the completion F_v. Then $\psi_\mathbb{A}(a) = \prod_v \psi_v(a_v)$ when $a = (a_v)$ is an adele, with $a_v \in F_v$ for every place v of F. It follows from the continuity of $\psi_\mathbb{A}$ that ψ_v is trivial on the ring $\mathfrak{o}_v$ of integers of F_v but no larger fractional ideal for all but finitely many v. Define a character ψ of $U(\mathbb{A})$ by (4.1), with ψ_F replaced by $\psi_\mathbb{A}$.

Now define

$$W_f(g) = \int_{U(\mathbb{A})} f(w_0 u g)\, \psi(u)^{-1} du.$$

THEOREM 4.3 *In the region of absolute convergence of the Eisenstein series, we have*

$$\int_{U(F)\backslash U(\mathbb{A})} E_\zeta(ug, f_\zeta)\psi(u)^{-1}\, du = W_f(g). \tag{4.4}$$

Proof. Using the Bruhat decomposition, representatives for $B(F)\backslash G(F)$ may be taken to be the set of $w^{-1}v$ where $w \in W$ and for each w, v runs through a set of representatives for

$$(U(F) \cap wU(F)w^{-1})\backslash U(F).$$

Therefore the left-hand side in (4.4) equals

$$\sum_{w\in W} \int_{(U(F)\cap wU(F)w^{-1})\backslash U(\mathbb{A})} f_\zeta(w^{-1}ug)\,\psi(u)^{-1}du =$$

$$\sum_{w\in W} \int_{(U(\mathbb{A})\cap wU(\mathbb{A})w^{-1})\backslash U(\mathbb{A})} \int_{(U(F)\cap wU(F)w^{-1})\backslash (U(\mathbb{A})\cap wU(\mathbb{A})w^{-1})} \psi(u')du'$$

$$f_\zeta(w^{-1}ug)\psi(u)^{-1}\, du.$$

If $w \neq w_0$ the inner integral is 0. On the other hand, if $w = w_0$ then the inner integral is trivial and may be discarded, and the statement follows. $\quad\square$

If $f \in V_\zeta$ is a pure tensor, this integral decomposes into a product over all places, so

$$\int_{U(\boldsymbol{F})\backslash U(\mathbb{A})} E_\zeta(ug, f_\zeta)\psi(u)^{-1}\, du = \prod_v W_{f_v}(g) \tag{4.5}$$

where W_{f_v} is given by the Casselman-Shalika formula (4.2) for almost all v. In the next chapter, we will interpret this as the p-part in a Weyl group multiple Dirichlet series. What is out of the scope of this book, but known due to the work of Brubaker, Bump and Friedberg [13] and McNamara [65] is that even when $n > 1$, the Weyl group multiple Dirichlet series $\boldsymbol{Z}(\boldsymbol{s}; \boldsymbol{m})$ are still Whittaker coefficients of Eisenstein series. The passage from global to local needed to apply McNamara's local results in this context will be discussed in a paper of Friedberg and McNamara.

Chapter Five

Tokuyama's Theorem

Let $z = \mathrm{diag}(z_1, \cdots, z_{r+1})$ be an element of the group $\hat{T}(\mathbb{C})$, which is the diagonal subgroup of $\mathrm{GL}_{r+1}(\mathbb{C})$. In the application to the Casselman-Shalika formula we will take the z_i to be the Langlands parameters. (In terms of the s_i the z_i are determined by the conditions that $\prod z_i = 1$ and $z_i/z_{i+1} = q^{1-2s_{r+1-i}}$.)

Let us write the Weyl character formula in the form

$$\left[\prod_{\alpha \in \Phi^+}(1 - z^\alpha)\right] s_\lambda(z) = z^\rho \sum_{w \in W}(-1)^{l(w_0 w)} z^{w(\lambda + \rho)}, \tag{5.1}$$

where we recall that $s_\lambda(z) = s_\lambda(z_1, \cdots, z_{r+1})$ is the Schur polynomial. The sum over the Weyl group on the right-hand side is the *numerator* in the Weyl formula and the product on the left is essentially the *Weyl denominator*. The Weyl vector ρ, we recall, is $\rho = (r, r-1, \cdots, 2, 1, 0)$.

We have seen in (4.4) and (4.2) that p-part of the Whittaker coefficient of the Eisenstein series is

$$\left[\prod_{\alpha \in \Phi^+}(1 - q^{-1}z^\alpha)\right] s_\lambda(z). \tag{5.2}$$

The similarity of this expression to (5.1) indicates that the p-parts of Whittaker coefficients of Eisenstein series can be profitably regarded as *a deformation of the numerator in the Weyl character formula*.

Thus we are led to deformations of the Weyl character formula. Tokuyama [70] gave such a deformation. It is an expression of $s_\lambda(z)$ as a ratio of a numerator to a denominator. The denominator is a deformation of the Weyl denominator, and the numerator is a sum over Gelfand-Tsetlin patterns with top row $\lambda + \rho$. It can be rewritten as a sum over $\mathcal{B}_{\lambda + \rho}$.

Tokuyama's formula has a parameter t, which can be specialized in various ways. If $t = 1$, it is the Weyl character formula. If $t = 0$, it is equivalent to the combinatorial definition of the Schur polynomial as a sum over semistandard Young tableaux, that is, over $\mathcal{B}_\lambda$. If t is specialized to $-q^{-1}$ then the denominator in Tokuyama's formula matches the product in (5.2). In this specialization, the numerator in Tokuyama's formula agrees with the p-parts of the Weyl group multiple Dirichlet series in the special case $n = 1$.

We will state and prove Tokuyama's formula, then translate it into the language of crystals.

If $\mathfrak{T}$ is a Gelfand-Tsetlin pattern in the notation (1.5) let $s(\mathfrak{T})$ be the number of entries a_{ij} with $i > 0$ such that $a_{i-1,j-1} < a_{ij} < a_{i-1,j}$. Let $l(\mathfrak{T})$ be the number of entries a_{ij} with $i > 0$ such that $a_{ij} = a_{i-1,j-1}$. Thus $l(\mathfrak{T})$ is the number of boxed elements in $\Gamma(\mathfrak{T})$, and $s(\mathfrak{T})$ is the number of entries that are neither boxed nor circled. For $0 \leqslant i \leqslant r$ let $d_i = d_i(\mathfrak{T}) = \sum_{j=i}^{r} a_{ij}$ be the i-th row sum of $\mathfrak{T}$.

We recall that the Schur polynomial s_λ is symmetric, and if z_i are the eigenvalues of $g \in \mathrm{GL}_{r+1}(\mathbb{C})$ then $s_\lambda(z_1, \cdots, z_{r+1}) = \chi_\lambda(g)$, where χ_λ is the character of the irreducible representation with highest weight λ. In this chapter there will be an induction on r, so we will sometimes write $\rho = \rho_r = (r, r-1, \cdots, 0)$. Let $\mathrm{GT}(\lambda) = \mathrm{GT}_r(\lambda)$ be the set of Gelfand-Tsetlin patterns with $r+1$ rows having top row λ.

THEOREM 5.1 (Tokuyama [70]) *We have*

$$\sum_{\substack{\mathfrak{T} \in \mathrm{GT}(\lambda+\rho) \\ \mathfrak{T} \text{ strict}}} (t+1)^{s(\mathfrak{T})} t^{l(\mathfrak{T})} z_1^{d_r} z_2^{d_{r-1}-d_r} \cdots z_{r+1}^{d_0-d_1} =$$

$$\left\{ \prod_{i<j} (z_j + z_i t) \right\} s_\lambda(z_1, \cdots, z_{r+1}). \tag{5.3}$$

For comparison with the original paper, we note that we have reversed the order of the z_i, which does not affect the Schur polynomial since it is symmetric. The following proof is essentially Tokuyama's original one.

Proof. Let $\Lambda^{(r)}$ be the ring of symmetric polynomials in $z_1, \cdots, z_r$. There is a homomorphism $\Lambda^{(r+1)} \longrightarrow \Lambda^{(r)}$ in which one sets $z_{r+1} \longmapsto 1$. The homomorphism is not injective, but its restriction to the homogeneous part of Λ of fixed degree r is injective. We note that as polynomials in the z_i both sides of (5.3) are homogeneous of degree $d_0 = |\lambda| + \frac{1}{2} r(r+1)$. (If λ is a partition then $|\lambda| = \sum_i \lambda_i$. See Macdonald [64] for background on partitions and symmetric functions.) Two homogeneous polynomials of the same degree are equal if they are equal when $z_{r+1} = 1$, so it is sufficient to show that (5.3) is true when $z_{r+1} = 1$, and for this we may assume inductively that the formula is true for $r - 1$.

The branching rule from $\mathrm{GL}_{r+1}(\mathbb{C})$ to $\mathrm{GL}_r(\mathbb{C})$, already mentioned in connection with (2.4), may be expressed as

$$s_\lambda(z_1, \cdots, z_r, 1) = \sum_{\mu \text{ interleaves } \lambda} s_\mu(z_1, \cdots, z_r). \tag{5.4}$$

See for example Bump [19], Chapter 44.

Also on setting $z_{r+1} = 1$,

$$\left\{ \prod_{1 \leqslant i < j \leqslant r+1} (z_j + z_i t) \right\}$$

becomes

$$\left\{ \prod_{1 \leqslant i < j \leqslant r} (z_j + z_i t) \right\} \left[\sum_{k=0}^{r} t^k e_k(z_1, \cdots, z_r) \right],$$

where e_k is the k-th elementary symmetric polynomial, that is, the sum of all squarefree monomials of degree k.

Now we recall Pieri's formula in the form

$$e_k s_\mu = \sum_{\substack{\nu \perp |\mu|+k \\ \nu\backslash\mu \text{ is a vertical strip}}} s_\nu.$$

See for example Bump [19], Chapter 42. The notation $\nu \perp |\mu| + k$ means that ν is a partition of $|\mu| + k$. The condition that $\nu\backslash\mu$ is a vertical strip means that the Young diagram of ν contains the Young diagram of μ, and that the skew-diagram $\nu\backslash\mu$ has no two entries in the same row. Thus $\nu\backslash\mu$ is a vertical strip if and only if each $\nu_i = \mu_i$ or $\mu_i + 1$, and since $\nu \perp |\mu| + k$, it follows that $\nu_i = \mu_i + 1$ exactly k times. Thus when $z_{r+1} = 1$ the right-hand side of (5.3) becomes

$$\sum_k \left\{ \prod_{1 \leqslant i < j \leqslant r} (z_j + z_i t) \right\} t^k \sum_{\mu \text{ interleaves } \lambda} \sum_{\substack{\nu \perp |\mu|+k \\ \nu\backslash\mu \text{ is a vertical strip}}} s_\nu(z_1, \cdots, z_r).$$

Now by induction

$$\left\{ \prod_{1 \leqslant i < j \leqslant r} (z_j + z_i t) \right\} s_\nu(z_1, \cdots, z_r) =$$
$$\sum_{\substack{\mathfrak{T}' \in \mathrm{GT}_{r-1}(\nu+\rho_{r-1}) \\ \mathfrak{T}' \text{ strict}}} (t+1)^{s(\mathfrak{T}')} t^{l(\mathfrak{T}')} z_1^{d'_{r-1}} z_2^{d'_{r-2}-d'_{r-1}} \cdots z_r^{d'_0 - d'_1},$$

where d'_i are the row sums of $\mathfrak{T}'$. We substitute this and interchange the order of summation and make the summation over μ the innermost sum. The condition that $\nu\backslash\mu$ is a vertical strip means that $\mu_i \leqslant \nu_i \leqslant \mu_i + 1$. Combining this with the fact that μ interleaves λ we have

$$\lambda_i + 1 \geqslant \mu_i + 1 \geqslant \nu_i \geqslant \mu_i \geqslant \lambda_{i+1} \tag{5.5}$$

and therefore $\nu + \rho_{r-1}$ interleaves $\lambda + \rho_r$. Since $k = |\nu| - |\mu|$, the right-hand side of (5.3), with z_{r+1} specialized to 1, equals

$$\sum_{\substack{\nu \\ \nu+\rho_{r-1} \text{ interleaves } \lambda + \rho_r}} \sum_{\substack{\mathfrak{T}' \in \mathrm{GT}_{r-1}(\nu+\rho_{r-1}) \\ \mathfrak{T}' \text{ strict}}} (t+1)^{s(\mathfrak{T}')} t^{l(\mathfrak{T}')}$$

$$\left[\sum_{\substack{\mu \text{ interleaving } \lambda \\ \nu \supset \mu \\ \nu\backslash\mu \text{ is a vertical strip}}} t^{|\nu|-|\mu|} \right] z_1^{d'_{r-1}} z_2^{d'_{r-2}-d'_{r-1}} \cdots z_r^{d'_0 - d'_1}.$$

Now we assemble $\lambda + \rho_r$ and the Gelfand-Tsetlin pattern $\mathfrak{T}'$ into a big Gelfand-Tsetlin pattern $\mathfrak{T}$. The row sums of $\mathfrak{T}$ and $\mathfrak{T}'$ are the same except the top row, so $d_i = d'_{i-1}$ for $1 \leqslant i \leqslant r$. We may just as well sum over $\mathfrak{T}$, in which case $\mathfrak{T}'$ is the pattern obtained by discarding the top row of $\mathfrak{T}$. We get

$$\sum_{\substack{\mathfrak{T}\in\mathrm{GT}_r(\lambda+\rho)\\ \mathfrak{T}\text{ strict}}} (t+1)^{s(\mathfrak{T}')} t^{l(\mathfrak{T}')} \left[\sum_{\substack{\mu\text{ interleaving }\lambda\\ \nu\supset\mu\\ \nu\backslash\mu\text{ is a vertical strip}}} t^{|\nu|-|\mu|} \right] z_1^{d_r}\cdots z_r^{d_1-d_2}.$$

We evaluate the term in brackets. It is

$$\prod_{i=1}^r \left[\sum_{\substack{\lambda_i\geqslant\mu_i\geqslant\lambda_{i+1}\\ \mu_i=\nu_i\text{ or }\nu_i-1}} t^{\nu_i-\mu_i} \right].$$

We now show that this equals $(t+1)^{s(\mathfrak{T})-s(\mathfrak{T}')} t^{l(\mathfrak{T})-l(\mathfrak{T}')}$. By (5.5), if $\nu_i=\lambda_i+1$ then $\nu_i=\mu_i+1$ and so $t^{\nu_i-\mu_i}=t$. The number of such terms is $l(\mathfrak{T})-l(\mathfrak{T}')$ and so we have a contribution of $t^{l(\mathfrak{T})-l(\mathfrak{T}')}$. If $\nu_i=\lambda_{i+1}$, then (5.5) implies that $\nu_i=\mu_i$ and these factors equal 1. Hence they may be discarded from the product. In the remaining cases we have $\lambda_i+1>\nu_i>\lambda_{i+1}$ both t and 1 can occur. The number of such terms is $s(\mathfrak{T})-s(\mathfrak{T}')$, and so we have a contribution of $(t+1)^{s(\mathfrak{T})-s(\mathfrak{T}')}$. Hence the term in brackets equals $t^{l_{\mathrm{top}}(\mathfrak{T})}(t+1)^{s_{\mathrm{top}}(\mathfrak{T})}$, where $l_{\mathrm{top}}(\mathfrak{T})$ is the number of boxed entries in the top row of $\Gamma(\mathfrak{T})$ and $s_{\mathrm{top}}(\mathfrak{T})$ is the number of entries in the top row of $\Gamma(\mathfrak{T})$ that are neither boxed nor circled. Clearly $l(\mathfrak{T})=l(\mathfrak{T}')+l_{\mathrm{top}}(\mathfrak{T})$ and $s(\mathfrak{T})=s(\mathfrak{T}')+s_{\mathrm{top}}(\mathfrak{T})$ so we obtain the left-hand side of (5.3), with $z_{r+1}=1$. This completes the induction. $\square$

We will give a version of Tokuyama's formula for crystals. We will take $n=1$ in (1.15). In this case the Gauss sums may be evaluated explicitly and

$$g(a)=q^a g^\flat, \qquad h(a)=q^a h^\flat$$

where

$$g^\flat=g^\flat(a)=-q^{-1}, \qquad h^\flat=h^\flat(a)=(q-1)q^{-1}$$

are independent of a.

LEMMA 5.2 *If $v\in\mathcal{B}_{\lambda+\rho}$ then*

$$G_\Gamma^\flat(v)=\begin{cases} (-q^{-1})^{l(\mathfrak{T}_v)}(1-q^{-1})^{s(\mathfrak{T}_v)} & \text{if }\mathfrak{T}_v\text{ is strict}\\ 0 & \text{otherwise.} \end{cases}$$

Proof. If $\mathfrak{T}_v$ is non-strict then some $a_{i,j}=a_{i,j+1}$. This means that

$$a_{i,j}=a_{i+1,j+1}=a_{i,j+1}$$

and so the entry $\Gamma_{i+1,j+1}$ is both boxed and circled in $\Gamma(\mathfrak{T}_v)$, which implies that $G_\Gamma^\flat(v)=G_\Gamma^\flat(\mathfrak{T}_v)=0$. On the other hand if $\mathfrak{T}_v$ is strict then

$$G_\Gamma^\flat(v)=\prod_i\begin{cases} -q^{-1} & \text{if }b_i\text{ is boxed,}\\ 1 & \text{if }b_i\text{ is circled,}\\ (q-1)q^{-1} & \text{if }b_i\text{ is neither boxed nor circled,} \end{cases}$$

which clearly equals $(-q^{-1})^{l(\mathfrak{T}_v)}(1-q^{-1})^{s(\mathfrak{T}_v)}$ and the statement is proved. $\square$

THEOREM 5.3 *If λ is a dominant weight, and if $z_1, \cdots, z_{r+1}$ are the eigenvalues of $g \in \mathrm{GL}_{r+1}(\mathbb{C})$, then*

$$\prod_{\alpha \in \Phi^+} (1 - q^{-1} \mathbf{z}^\alpha) \chi_\lambda(g) = \sum_{v \in \mathcal{B}_{\lambda+\rho}} G_\Gamma^\flat(v) \mathbf{z}^{\mathrm{wt}(v) - w_0 \rho}.$$

Proof. We will prove this in the form

$$\sum_{v \in \mathcal{B}_{\lambda+\rho}} G_\Gamma^\flat(v) \mathbf{z}^{\mathrm{wt}(v)} = \prod_{i<j} (z_j - q^{-1} z_i) \chi_\lambda(g). \tag{5.6}$$

This is equivalent since

$$\mathbf{z}^{w_0 \rho} \prod_{\alpha \in \Phi^+} (1 - q^{-1} \mathbf{z}^\alpha) = z_2 z_3^2 \cdots z_{r+1}^r \prod_{i<j} \left(1 - q^{-1} \frac{z_i}{z_j} \right).$$

By Theorem 5.1 with $t = -q^{-1}$, the right-hand side of (5.6) equals

$$\sum_{\substack{\mathfrak{T} \in \mathrm{GT}(\lambda+\rho) \\ \mathfrak{T} \text{ strict}}} (1 - q^{-1})^{s(\mathfrak{T})} (-q^{-1})^{l(\mathfrak{T})} z_1^{d_r} z_2^{d_{r-1}-d_r} \cdots z_{r+1}^{d_0-d_1}.$$

Turning to the left-hand side of (5.6), let $v \in \mathcal{B}_{\lambda+\rho}$. By (2.17) we have $\mathbf{z}^{\mathrm{wt}(v)} = z_1^{d_r} z_2^{d_{r-1}-d_r} \cdots z_{r+1}^{d_0-d_1}$. Combining this with Lemma 5.2, we obtain (5.6). $\qquad \square$

Chapter Six

Outline of the Proof

The proof of Theorem 1.1 involves many remarkable phenomena, and we wish to explain its structure in this chapter. To this end, we will give the first of a succession of statements, each of which implies the theorem. Passing from each statement to the next is a nontrivial reduction that changes the nature of the problem to be solved. We will outline the ideas of these reductions here and tackle them in detail in subsequent chapters.

Statement A. *We have* $H_\Gamma = H_\Delta$.

This reduction was already mentioned in the first chapter, where Statement A appeared as Theorem 1.2.

The proof that this implies Theorem 1.1 is Theorem 1 of [15]. We review the idea of the proof. To prove the functional equations that $Z_\Psi(s; m)$ is to satisfy, using the method described in [20], [10], [12], and [14] based on Bochner's convexity principle [7], one must prove meromorphic continuation to a larger region and a functional equation for each generator $\sigma_1, \cdots, \sigma_r$ of the Weyl group – the simple reflections. These act on the coordinates by

$$
\sigma_i(s_j) = \begin{cases} 1 - s_j & \text{if } j = i, \\ s_i + s_j - \frac{1}{2} & \text{if } j = i \pm 1, \\ s_j & \text{if } |j - i| > 1. \end{cases}
$$

We proceed inductively. Taking $H = H_\Gamma$ as the definition of the series, all but one of these functional equations may be obtained by collecting the terms to produce a series whose terms are multiple Dirichlet series of lower rank. To see this reduction, note that we have described the p-part of H as a sum over Gelfand-Tsetlin patterns, extended this to a definition to $H(c_1, \cdots, c_r; m_1, \cdots, m_r)$ by (twisted) multiplicativity. Equivalently, one may define $H(c_1, \cdots, c_r; m_1, \cdots, m_r)$ by specifying a Gelfand-Tsetlin pattern $\mathfrak{T}_p$ for each prime; for all but finitely many p the pattern must be the minimal one

$$
\left\{ \begin{matrix} r & & r-1 & & \cdots & & 0 \\ & r-1 & & & \cdots & & 0 \\ & & \ddots & & & \iddots & \\ & & & 0 & & & \end{matrix} \right\}.
$$

Summing over patterns with fixed top row (determined by the $\mathrm{ord}_p(m_i)$) and fixed row sums (determined by $\mathrm{ord}_p(c_i)$) gives $H(c_1, \cdots, c_r; m_1, \cdots, m_r)$. More precisely we may group the terms as follows. For each prime p of $\mathfrak{o}_S$, fix a partition λ_p of length r into unequal parts such that for almost all p we have $\lambda_p =$

$(r, r - 1, \cdots, 0)$; then collect the terms in which for each p the top row of $\mathfrak{T}_p$ is λ_p. These produce an exponential factor times a term $Z(\boldsymbol{s}; \boldsymbol{m}'; A_{r-1})$ where $\boldsymbol{m}'$ depends on λ_p (for each p). This expansion gives, by induction, the functional equations for the subgroup of W generated by $\sigma_2, \cdots, \sigma_r$. Similarly starting with $H = H_\Delta$ gives functional equations for the subgroup generated by $\sigma_1, \cdots, \sigma_{r-1}$. Notice that these two sets of reflections generate all of W. If Statement A holds, then combining these analytic continuations and functional equations and invoking Bochner's convexity principle gives the required analytic continuation and functional equations. We refer to [15] for further details.

Since H_Γ and H_Δ satisfy the same twisted multiplicativity, it suffices to work at powers of a single prime p. We see that there are two ways in which these coefficients differ. First, given a lattice point $\boldsymbol{k} = (k_1, \ldots, k_r)$ in the polytope defined by the Gelfand-Tsetlin patterns of given top row, there are two ways of attaching a set of Gelfand-Tsetlin patterns to $\boldsymbol{k}$, namely the set of $\mathfrak{T}$ with $k_\Gamma(\mathfrak{T}) = \boldsymbol{k}$, or with $k_\Delta(\mathfrak{T}) = \boldsymbol{k}$. Second, given a pattern $\mathfrak{T}$, there are two ways of attaching numbers to it: $G_\Gamma(\mathfrak{T})$, resp. $G_\Delta(\mathfrak{T})$.

An attack on Statement A can be formulated using the *Schützenberger involution* on Gelfand-Tsetlin patterns. This is the involution q_r, which was already defined in Chapter 1. It interchanges the functions k_Δ and k_Γ, and one may thus formulate Statement A as saying that

$$\sum_{k_\Gamma(\mathfrak{T}) = (k_1, \cdots, k_r)} G_\Gamma(\mathfrak{T}) = \sum_{k_\Gamma(\mathfrak{T}) = (k_1, \cdots, k_r)} G_\Delta(q_r \mathfrak{T}). \tag{6.1}$$

Remark. In many cases, for example if $\mathfrak{T}$ is in the interior of the polytope of Gelfand-Tsetlin patterns with fixed $k_\Gamma(\mathfrak{T})$, it may be shown that

$$G_\Gamma(\mathfrak{T}) = G_\Delta(q_r \mathfrak{T}).$$

If this were always true there would be no need to sum in (6.1). In general, however, this is false. What *is* ultimately true is that the patterns may be partitioned into fairly small "packets" such that if one sums over a packet, $\sum G_\Gamma(\mathfrak{T}) = \sum G_\Delta(q_r \mathfrak{T})$. The packets, we observe, can be identified empirically in any given case, but are difficult to characterize in general, and not even uniquely determined in some cases. We will return to the matter of packets later.

To proceed further, we introduce the notion of a *short Gelfand-Tsetlin pattern* or (more briefly) a *short pattern*. By this we mean an array with just three rows

$$\mathfrak{t} = \left\{ \begin{array}{ccccccc} l_0 & & l_1 & & l_2 & \cdots & l_{d+1} \\ & a_0 & & a_1 & & a_d & \\ & & b_0 & & \cdots & b_{d-1} & \end{array} \right\} \tag{6.2}$$

where the rows are nonincreasing sequences of integers that interleave, that is,

$$l_i \geqslant a_i \geqslant l_{i+1}, \qquad a_i \geqslant b_i \geqslant a_{i+1}. \tag{6.3}$$

We will refer to $l_0, \cdots, l_{d+1}$ as the *top* or *zero-th row* of $\mathfrak{t}$, $a_0, \cdots, a_d$ as the *first* or *middle row* and $b_0, \cdots, b_{d-1}$ as the *second* or *bottom row*. We may assume that the top and bottom rows are strict, but we need to allow the first row to be non-strict. We define the *weight k* of $\mathfrak{t}$ to be the sum of the a_i.

If t is a short pattern we define another short pattern

$$t' = \left\{ \begin{array}{ccccccc} l_0 & & l_1 & & l_2 & \cdots & & l_{d+1} \\ & a_0' & & a_1' & & & a_d' & \\ & & b_0 & & \cdots & b_{d-1} & & \end{array} \right\}, \qquad (6.4)$$

where

$$a_j' = \min(l_j, b_{j-1}) + \max(l_{j+1}, b_j) - a_j, \qquad 0 < j < d, \qquad (6.5)$$

$$a_0' = l_0 + \max(l_1, b_0) - a_0, \qquad a_d' = \min(l_d, b_{d-1}) + l_{d+1} - a_d. \qquad (6.6)$$

We call t' the (Schützenberger) *involute* of t. To see why this definition is reasonable, note that if the top and bottom rows of t are specified, then a_i are constrained by the inequalities

$$\min(l_j, b_{j-1}) \geqslant a_j \geqslant \max(l_{j+1}, b_j), \qquad 0 < j < d, \qquad (6.7)$$

$$l_0 \geqslant a_0 \geqslant \max(l_1, b_0), \qquad \min(l_d, b_{d-1}) \geqslant a_d \geqslant l_{d+1}. \qquad (6.8)$$

These inequalities express the assumption that the three rows of the short pattern interleave. The array t' is obtained by reflecting a_j in its permitted range.

The Schützenberger involution of full Gelfand-Tsetlin patterns is built up from operations involving three rows at a time, based on the operation $t \longmapsto t'$ of short Gelfand-Tsetlin patterns. This is done $\frac{1}{2}r(r+1)$ times to obtain the Schützenberger involution. Using this decomposition and induction, we prove that to establish Statement A one needs only the equivalence of two sums of Gauss sums corresponding to Gelfand-Tsetlin patterns that differ by a single involution. This allows us to restrict our attention, within Gelfand-Tsetlin patterns, to short patterns. To be more precise and to explain what must be proved, we make the following definitions.

By a short pattern *prototype* $\mathfrak{S}$ of length d we mean a triple $(\boldsymbol{l}, \boldsymbol{b}, k)$ specifying the following data: a top row consisting of an integer sequence $\boldsymbol{l} = (l_0, \cdots, l_{d+1})$, a bottom row consisting of a sequence $\boldsymbol{b} = (b_0, \cdots, b_d)$, and a positive integer k. It is assumed that $l_0 > l_1 > \ldots > l_{d+1}$, that $b_0 > b_1 > \ldots > b_{d-1}$ and there is no loss of generality in assuming that $l_i \geqslant b_i \geqslant l_{i+2}$.

We say that a short pattern t of length d *belongs to the prototype* $\mathfrak{S}$ if it has the prescribed top and bottom rows, and its weight is k (so $\sum_i a_i = k$). By abuse of notation, we will use the notation $t \in \mathfrak{S}$ to mean that t belongs to the prototype $\mathfrak{S}$. Prototypes were called *types* in [15], but we will reserve that term for a more restricted equivalence class of short patterns.

Given a short Gelfand-Tsetlin pattern, we may define two two-rowed arrays Γ_t and Δ_t, to be called *preaccordions*, which display information used in the evaluations we must make. These are defined analogously to the arrays associated with a full Gelfand-Tsetlin pattern, which were denoted $\Gamma(t)$ and $\Delta(t)$. There is an important distinction in that in Γ_t we use the right-hand rule on the first row, and the left-hand rule on the second row, and for Δ_t we reverse these. In the array $\Gamma(\mathfrak{T})$ for the full Gelfand-Tsetlin pattern we used the right-hand rule for every row, and in $\Delta(\mathfrak{T})$ we used the left-hand rule for every row.

Thus

$$\Gamma_t = \left\{ \begin{array}{ccccc} \mu_0 & \mu_1 & \cdots & & \mu_d \\ & \nu_0 & & \cdots & \nu_{d-1} \end{array} \right\}, \tag{6.9}$$

and

$$\Delta_t = \left\{ \begin{array}{ccccc} \kappa_0 & \kappa_1 & \cdots & & \kappa_d \\ & \lambda_0 & & \cdots & \lambda_{d-1} \end{array} \right\}, \tag{6.10}$$

where

$$\mu_j = \sum_{k=j}^{d}(a_k - l_{k+1}), \qquad \nu_j = \sum_{k=0}^{j}(a_k - b_k),$$

and

$$\kappa_j = \sum_{k=0}^{j}(l_k - a_k), \qquad \lambda_j = \sum_{k=j}^{d-1}(b_k - a_{k+1}).$$

We also use the right-hand rule to describe the circling and boxing of the elements of the first row of Γ_t, and the left-hand rule to describe the circling and boxing of elements of the bottom row, reversing these for Δ_t. This means we circle μ_j if $a_j = l_{j+1}$ and box μ_i if $a_j = l_j$; we circle ν_j if $b_j = a_j$ and box ν_j if $b_j = a_{j+1}$. The boxing and circling rules are reversed for Δ_t: we box κ_j if $\alpha_j = l_{j+1}$ and circle α_i if $\alpha_j = l_j$; we box λ_j if $b_j = a_j$ and box λ_j if $b_j = a_{j+1}$.

We give an example to illustrate these definitions. Suppose that

$$t = \left\{ \begin{array}{ccccccc} 23 & & 15 & & 12 & & 5 & & 2 & & 0 \\ & 20 & & 12 & & 5 & & 4 & & 2 \\ & 14 & & 9 & & 5 & & 3 \end{array} \right\}.$$

Then

$$\Gamma_t = \left\{ \begin{array}{ccccc} 9 & \boxed{\textcircled{4}} & \textcircled{4} & 4 & \boxed{2} \\ & 6 & 9 & \textcircled{9} & 10 \end{array} \right\}.$$

We have indicated the circling and boxing of entries. Now applying the involution,

$$t' = \left\{ \begin{array}{ccccccc} 23 & & 15 & & 12 & & 5 & & 2 & & 0 \\ & 18 & & 14 & & 9 & & 4 & & 0 \\ & 14 & & 9 & & 5 & & 3 \end{array} \right\},$$

and

$$\Delta_{t'} = \left\{ \begin{array}{ccccc} 5 & 6 & 9 & 10 & \boxed{12} \\ & \textcircled{4} & \textcircled{4} & 4 & 3 \end{array} \right\}.$$

We observe the following points.

- The first row of Γ_t is decreasing and the bottom row is increasing; these are reversed for $\Delta_{t'}$, just as the boxing and circling conventions are reversed.

- The involution does not preserve strictness. If t is strict, no element can be both boxed and circled, but if t is not strict, an entry in the bottom row is both boxed and circled, and the same is true for $\Delta_{t'}$: if t' is not strict, then an entry in the bottom row of $\Delta_{t'}$ is both boxed and circled.

- It might be more natural to use the notations $\Gamma\Delta_t$ and $\Delta\Gamma_t$, and to refer to these as "$\Gamma\Delta$-accordions" and "$\Delta\Gamma$-accordions," to indicate that each has a Gamma layer (using the right-hand rule) and a Delta layer (using the left-hand rule). We will not do this since that would be notationally cumbersome.

 If $\mathfrak{T}$ is a Gelfand-Tsetlin pattern, then in $\Gamma(\mathfrak{T})$ we use the right-hand rule in every row, and in $\Delta(\mathfrak{T})$ we use the left-hand row in every row. But if t is a *short* Gelfand-Tsetlin pattern then in Γ_t and Δ_t one row uses the right-hand rule, the other the left-hand rule.

Let us define

$$
G_\Gamma(t) = \prod_{x \in \Gamma_t} \begin{cases} g(x) & \text{if } x \text{ is boxed in } \Gamma_t \text{ but not circled;} \\ q^x & \text{if } x \text{ is circled but not boxed;} \\ h(x) & \text{if } x \text{ is neither boxed nor circled;} \\ 0 & \text{if } x \text{ is both boxed and circled.} \end{cases}
$$

Thus if t is not strict, then $G_\Gamma(t) = 0$. Similarly, let

$$
G_\Delta(t') = \prod_{x \in \Delta_{t'}} \begin{cases} g(x) & \text{if } x \text{ is boxed in } \Delta_{t'} \text{ but not circled;} \\ q^x & \text{if } x \text{ is circled but not boxed;} \\ h(x) & \text{if } x \text{ is neither boxed nor circled;} \\ 0 & \text{if } x \text{ is both boxed and circled.} \end{cases}
$$

We also have reduced versions

$$
G_\Gamma^\flat(t) = \prod_{x \in \Gamma_t} \begin{cases} g^\flat(x) & \text{if } x \text{ is boxed in } \Gamma_t \text{ but not circled;} \\ 1 & \text{if } x \text{ is circled but not boxed;} \\ h^\flat(x) & \text{if } x \text{ is neither boxed nor circled;} \\ 0 & \text{if } x \text{ is both boxed and circled.} \end{cases}
$$

$$
G_\Delta^\flat(t') = \prod_{x \in \Delta_{t'}} \begin{cases} g^\flat(x) & \text{if } x \text{ is boxed in } \Delta_{t'} \text{ but not circled;} \\ 1 & \text{if } x \text{ is circled but not boxed;} \\ h^\flat(x) & \text{if } x \text{ is neither boxed nor circled;} \\ 0 & \text{if } x \text{ is both boxed and circled.} \end{cases}
$$

Thus in our examples,

$$
G_\Gamma(t) = h(9) \cdot q^4 \cdot q^4 \cdot h(4) \cdot g(2) \cdot h(6) \cdot h(9) \cdot q^9 \cdot h(10) = q^{57} G_\Gamma^\flat
$$

and

$$
G_\Delta(t') = h(5) \cdot h(6) \cdot h(9) \cdot h(12) \cdot g(12) \cdot q^4 \cdot q^4 \cdot h(4) \cdot h(3) = q^{57} G_\Delta^\flat.
$$

Statement B. *Let $\mathfrak{S}$ be a short pattern prototype. Then*

$$\sum_{t \in \mathfrak{S}} G_{\Gamma}(t) = \sum_{t \in \mathfrak{S}} G_{\Delta}(t'). \tag{6.11}$$

It is not hard to see that the sum M of the elements of Γ_t equals the sum of the elements of $\Delta_{t'}$, and

$$G_{\Gamma}(t) = q^M G_{\Gamma}^{\flat}(t), \qquad G_{\Delta}(t') = q^M G_{\Delta}^{\flat}(t').$$

Therefore Statement B is equivalent to the reduced form

$$\sum_{t \in \mathfrak{S}} G_{\Gamma}^{\flat}(t) = \sum_{t \in \mathfrak{S}} G_{\Delta}^{\flat}(t'). \tag{6.12}$$

Statement B was conjectured in [15]. A reinterpretation of Statement B in terms of crystal bases will be given in Chapter 18. Another reinterpretation in terms of lattice models in statistical mechanics is given in Chapter 19 and [9].

Just as in (6.1), the equality (6.12) is *almost* true term-by-term. Indeed, by Theorems 10.1 and 10.2 below we actually have, for many t (in some sense most)

$$G_{\Gamma}(t) = G_{\Delta}(t'). \tag{6.13}$$

For example if t is on the interior of the polytope of short patterns with fixed middle row sum, it can be proved that $G_{\Gamma}(t) = G_{\Delta}(t')$. This is precisely analogous to the situation with (6.1), and studying the phenomena that occur in connection with (6.11) gives us insight into (6.1).

The reduction to Statement B was proved in [15], which was written before Statement B was proved. We will repeat this argument (based on the Schützenberger involution) in Chapter 7. In a nutshell, Statement A can be deduced from Statement B because the Schützenberger involution q_r is built up from the involution $t \longmapsto t'$ of short Gelfand-Tsetlin patterns by repeated applications, and this will be explained in detail in Chapter 7.

Instead of pursuing the identification of packets as suggested in [15], we proceed by using (6.1) to reduce Statement A to Statement B, and the mysterious packets will eventually be sorted out by further combinatorial transformations of the problem that we will come to presently (Statements C, D, E, F, and G).

Most our effort will be devoted to the proof of Statement B, to which we now turn. We call the short pattern *totally resonant* if the bottom row repeats the top row, that is, if it has the form

$$t = \left\{ \begin{matrix} l_0 & & l_1 & & l_2 & & \cdots & & l_{d+1} \\ & a_0 & & a_1 & & & & a_d & \\ & & l_1 & & & \cdots & & l_d & \end{matrix} \right\}. \tag{6.14}$$

Notice that this is a property of the prototype, whose data consists of the top and bottom rows, and the middle row sum.

Statement C. *Statement B is true for totally resonant short pattern prototypes.*

The fact that Statement C implies Statement B will be proved in Theorem 13.7, which requires the introduction of some new concepts. Given a short pattern (6.2), we will associate with its prototype a certain graph called the *cartoon* by connecting

a_i to either l_i if $i = 0$, or else to either l_i or b_{i-1}, whichever is numerically closer to a_i. If $b_{i-1} = l_i$, then we connect a_i to both. We will also connect a_i to either l_{i+1} or b_i, whichever is numerically closer to a_i, or to both if they are equal, provided $i < d$; and a_d is connected to l_{d+1}. The connected components of the cartoon are called *episodes*.

We will subdivide the prototype into smaller equivalence classes called *types*. Two patterns t_1 and t_2 have the same type if, first of all, they have the same prototype (hence the same cartoon), and if for each episode $\mathcal{E}$, the sum of the a_i that lie in $\mathcal{E}$ are the same for t_1 and t_2. Statement B follows from the stronger statement that (6.11) holds when we sum over a type. We will reduce this statement to a series of separate problems, one for each episode. But each of these problems will be reduced to a single common problem (Statement D below) that is equivalent to Statement B for totally resonant prototypes.

The reduction to the totally resonant case involves some quite fascinating combinatorial phenomena. A key point is a combinatorial Lemma, which was called the "Snake Lemma" in [15], but which is not the familiar Snake Lemma from homological algebra. This Lemma says the elements in Γ_t can be matched up with the elements in $\Delta_{t'}$ in a bijection that has quite surprising properties. (The "snakes" appear in graphing this bijection.)

In addition to such combinatorial phenomena, number theory enters as well, in particular in the "Knowability Lemma" (Proposition 12.2), which we now briefly discuss. The term "knowability" refers to the fact that Gauss sums such as $g(a)$ when $n \nmid a$ have known absolute values, but their arguments as complex numbers are still mysterious. They are "unknowable." We will refer to an expression that is a product of terms of the form q^a, $h(a)$ and $g(a)$ as "knowable" if it can given a closed expression as a polynomial in q. Thus $g(a)$, taken in isolation, is unknowable unless $n|a$, in which case $g(a) = -q^{a-1}$. But even if $n \nmid a$ the product $g(a)g(b)$ is knowable if $n|a + b$ since then $g(a)g(b) = q^{a+b-1}$.

There is a strong tendency for the Gauss sums that appear in the terms $G_\Gamma(t)$ (for short patterns t) or in $G_\Gamma(\mathfrak{T})$ (for full Gelfand-Tsetlin patterns $\mathfrak{T}$), to appear in knowable combinations. The Knowability Lemma give an explanation for this. Moreover it gives key information that is needed for the sequel. Stable patterns are an important exception. We recall that a pattern is *stable* if every entry (except those in the top row) equals one of the two directly above it. The stable patterns are in a sense the most important ones, since they are the *only* patterns that contribute in the stable case (when n is large). If the Gelfand-Tsetlin patterns with fixed top row are embedded into a Euclidean space, the stable patterns are the extremal ones. The Gauss sums that appear in the stable terms are unknowable. Thus when $r = 1$, $Z_\Psi(s; m)$ is Kubota's Dirichlet series [53], and its use by Heath-Brown and Patterson [40] to study the distribution of cubic Gauss sums exploited precisely the appearance of such unknowable terms.

In order to prove Statement C, we must work with the evaluations of $G_\Gamma(t)$ and $G_\Delta(t')$. To do so, it is convenient to describe these evaluations by introducing some new notation and terminology. A Γ-*accordion* (of length d and weight s) is an array of nonnegative integers

$$\mathfrak{a} = \left\{ \begin{matrix} s & & \mu_1 & & \mu_2 & & \cdots & & \mu_d \\ & \nu_1 & & \nu_2 & & \cdots & & \nu_d & \end{matrix} \right\}, \qquad (6.15)$$

in which the first row is decreasing, the second increasing, and $\mu_i + \nu_i = s$. Thus if t is a short Gelfand-Tsetlin pattern, then the preaccordion Γ_t is an accordion if the condition $\mu_i + \nu_i = s$ is satisfied. We will sometimes write

$$\mu_0 = \nu_{d+1} = s, \qquad \nu_0 = \mu_{d+1} = 0. \qquad (6.16)$$

Also, by a Δ-*accordion* (of length $d + 1$ and weight s) we mean an array

$$\mathfrak{a}' = \left\{ \begin{matrix} \nu_1 & & \nu_2 & & \cdots & & \nu_d & & s \\ & \mu_1 & & \mu_2 & & \cdots & & \mu_d & \end{matrix} \right\}, \qquad (6.17)$$

where the first row is increasing, the bottom decreasing, and $\mu_i + \nu_i = s$. We will make use of the map $\mathfrak{a} \longmapsto \mathfrak{a}'$ that takes Γ-accordions to Δ-accordions.

The significance of these definitions is that if t is a totally resonant short Gelfand-Tsetlin pattern, then its Γ-preaccordion is a Γ-accordion. Moreover the Δ-pre-accordion of t′ is the Δ-accordion $\mathfrak{a}'$.

We have already described the decoration of preaccordions with boxes and circles. In the special case of an accordion, the decorations have some pleasant additional properties because they come from totally resonant short patterns.

- No entry of the first row is both boxed and circled. An entry of the bottom row may be both boxed and circled, in which case we say the accordion is *non-strict*.

- In a Γ-accordion, a bottom row entry is circled if and only if the entry above it and to the left is circled, and a bottom row entry is boxed if and only if the entry above it and to the right is boxed. Thus in (6.15), ν_i is circled if and only if the μ_{i-1} is circled, and ν_i is boxed if and only if μ_i is boxed.

- In a Δ-accordion, a bottom row entry is circled if and only if the entry above it and to the right is circled, and a bottom row entry is boxed if and only if the entry above it and to the left is boxed.

- In either (6.15) or (6.17) μ_i is circled if and only if $\mu_i = \mu_{i+1}$, and ν_i is circled if and only if $\nu_i = \nu_{i-1}$. (But note that μ_i is in the first row in (6.15) but in the second row in (6.17).) Invoking (6.16), special cases of this rule are that s is circled if and only if $s = \mu_1$, μ_d is circled if and only if $\mu_d = 0$, and ν_1 is circled if and only if $\nu_1 = 0$.

- There is no corresponding rule for the boxing. Thus the circling is determined by $\mathfrak{a}$ but the boxing is not.

We also note that the Knowability Lemma mentioned above allows us to assume that $n|s$.

Our goal is to systematically describe all decorated accordions that arise and their corresponding evaluations, so that we may sum over them and establish Statement

C. If $\mathfrak{a}$ is a decorated accordion (of either kind), then in view of the second and third rules, the decoration of the second row is determined by the decoration of the top row. We encode this by a *signature*, which is by definition a string $\sigma = \sigma_0 \cdots \sigma_d$, where each σ_i is one of the symbols $\bigcirc, \square$ or $*$. We associate a signature with a decorated accordion by taking $\sigma_i = \bigcirc$ if μ_i is circled in the first row (with $\mu_0 = s$, of course), $\square$ if μ_i is boxed, and $*$ if it is neither circled or boxed. We say the accordion $\mathfrak{a}$ and the signature σ are *compatible* if the following *circling compatibility condition* is satisfied (for conformity with the rules already stated for the decorations). For Γ-accordions labeled as in (6.15) the condition is

$$\sigma_i = \bigcirc \text{ if and only if } \mu_i = \mu_{i+1}. \tag{6.18}$$

In view of (6.16), if $i = 0$, this means $s = \mu_1$, and if $i = d$ it means $\mu_d = 0$. For Δ-accordions labeled as in (6.17) the condition is

$$\sigma_i = \bigcirc \text{ if and only if } \nu_i = \nu_{i-1},$$

which means that $\sigma_0 = \bigcirc$ if and only if $\nu_1 = 0$, and $\sigma_d = \bigcirc$ if and only if $s = \nu_1$.

Since the signature determines the decoration, we will denote by $\mathfrak{a}_\sigma$ the decorated accordion, where σ is a signature compatible with the accordion $\mathfrak{a}$. We will apply the same signature σ to the involute $\mathfrak{a}'$. Thus if $\sigma = * \square * * \bigcirc *$ and

$$\mathfrak{a} = \left\{ \begin{array}{cccccc} 9 & 7 & 6 & 4 & 2 & 2 \\ & 2 & 3 & 5 & 7 & 7 \end{array} \right\}$$

then

$$\mathfrak{a}_\sigma = \left\{ \begin{array}{cccccc} 9 & \boxed{7} & 6 & 5 & \textcircled{2} & 2 \\ & 2 & 3 & 5 & 7 & \textcircled{7} \end{array} \right\},$$

which is a decorated Γ-accordion, while

$$\mathfrak{a}'_\sigma = \left\{ \begin{array}{cccccc} 2 & \boxed{3} & 5 & 7 & \textcircled{7} & 9 \\ & 7 & \boxed{6} & 4 & \textcircled{2} & 2 \end{array} \right\},$$

which is a decorated Δ-accordion. Observe that the signature encodes the decoration of the first row, and since we use the same signature in both $\mathfrak{a}$ and $\mathfrak{a}'$, it follows that the location of the boxes and circles in the first row is the same for $\mathfrak{a}$ as for the Δ-accordion $\mathfrak{a}'$. However the decoration of the bottom rows are different. This is because the boxing and circling rules are different for Γ-accordions and Δ-accordions.

Now if $\mathfrak{a}_\sigma$ is a decorated Γ-accordion, let

$$\mathcal{G}_\Gamma(\mathfrak{a}, \sigma) = \mathcal{G}_\Gamma(\mathfrak{a}_\sigma) = \prod_{x \in \mathfrak{a}} \begin{cases} g(x) & \text{if } x \text{ is boxed in } \mathfrak{a}_\sigma \text{ but not circled,} \\ q^x & \text{if } x \text{ is circled but not boxed,} \\ h(x) & \text{if } x \text{ is neither boxed nor circled,} \\ 0 & \text{if } x \text{ is both boxed and circled.} \end{cases}$$

The notations $\mathcal{G}_\Gamma(\mathfrak{a}_\sigma)$ and $\mathcal{G}_\Gamma(\mathfrak{a}, \sigma)$ are synonyms; we will prefer the former when working with the free abelian group on the decorated accordions, the latter when $\mathfrak{a}$ is fixed and σ is allowed to vary.

If $\mathfrak{a}'_\sigma$ is a decorated Δ-accordion, we define $\mathcal{G}_\Delta(\mathfrak{a}'_\sigma)$ by the same formula. We retain the subscripts Γ and Δ since $\mathcal{G}_\Gamma$ and $\mathcal{G}_\Delta$ have different domains. Thus in the last example

$$\mathcal{G}_\Gamma(\mathfrak{a}_\sigma) = h(9)g(7)g(2)h(6)h(3)h(5)h(5)q^2h(7)h(2)q^7,$$

$$\mathcal{G}_\Delta(\mathfrak{a}'_\sigma) = h(2)h(7)g(3)g(6)h(5)h(4)h(7)q^2q^7h(2)h(9).$$

Now let positive integers $s, c_0, c_1, \cdots, c_d$ be given. By the Γ-*resotope* (of length d), to be denoted $\mathcal{A}^\Gamma_s(c_0, c_1, \cdots, c_d)$, we mean the sum, in the free abelian group $\mathfrak{Z}_\Gamma$ on the set of decorated accordions, of $\mathfrak{a}_\sigma$ such that the parameters in $\mathfrak{a}$ satisfy

$$0 \leqslant s - \mu_1 \leqslant c_0, \quad 0 \leqslant \mu_1 - \mu_2 \leqslant c_1, \quad \cdots \quad, 0 \leqslant \mu_d \leqslant c_d, \tag{6.19}$$

and

$$\sigma_i = \begin{cases} \circ & \text{if } \mu_i - \mu_{i+1} = 0; \\ \square & \text{if } \mu_i - \mu_{i+1} = c_i; \\ * & \text{if } 0 < \mu_i - \mu_{i+1} < c_i. \end{cases}$$

If $\mathcal{A} = \mathcal{A}^\Gamma_s(c_0, c_1, \cdots, c_d)$, by abuse of notation we may write $\mathfrak{a}_\sigma \in \mathcal{A}$ to mean that $\mathfrak{a}_\sigma$ appears with nonzero coefficient in $\mathcal{A}$ as described above. Let $\mathcal{A}'$ be the image of $\mathcal{A}$ under the involution $\mathfrak{a}_\sigma \longmapsto \mathfrak{a}'_\sigma$; we call $\mathcal{A}'$ a Δ-*resotope*.

The set of Γ-accordions of the form (6.15), embedded into Euclidean space by mapping $\mathfrak{a} \longmapsto (\mu_1, \cdots, \mu_d)$, may be regarded as the set of lattice points in a polytope. The resotopes we have just described correspond to these points with compatible decorations and signatures attached. We will sometimes discuss the geometry of the underlying polytope without making explicit note of the additional data attached to each point.

Given a totally resonant type, we will prove in the Corollary to Proposition 13.2 below that the accordion Γ_t runs through a resotope $\mathcal{A}$, and $\Delta_{t'}$ runs through $\mathcal{A}'$. This allows us to pass from Statement C to the following statement.

Statement D. *Let $\mathcal{A}$ be a Γ-resotope. Then*

$$\sum_{\mathfrak{a}_\sigma \in \mathcal{A}} \mathcal{G}_\Gamma(\mathfrak{a}_\sigma) = \sum_{\mathfrak{a}'_\sigma \in \mathcal{A}'} \mathcal{G}_\Delta(\mathfrak{a}'_\sigma). \tag{6.20}$$

We turn to the proof of Statement D. We will show that if $\mathfrak{a}$ lies in the interior of the resotope, then its signature is just $* \cdots *$, and we have $\mathcal{G}_\Gamma(\mathfrak{a}_\sigma) = h(s) \prod_{i=1}^d h(\mu_i)h(\nu_i) = \mathcal{G}_\Delta(\mathfrak{a}'_{\sigma'})$. This may fail, however, when $\mathfrak{a}$ is on the boundary, so this is the remaining obstacle to proving Statement D. The approach suggested in [15] is to try to partition the boundary into small "packets" such that the sums over each packet are equal. In practice one can carry this out in any given case, but giving a coherent theory of packets along these lines seems unpromising. First, the resotopes themselves are geometrically complex. Second, even when the resotope is geometrically simple, the identification of the packets can be perplexing, and devoid of any apparent pattern. We will give an example illustrating this approach at the end of Chapter 13.

Geometrically, a resotope is a figure obtained from a simplex by chopping off some of the corners; the pieces removed are themselves simplices. But the resulting

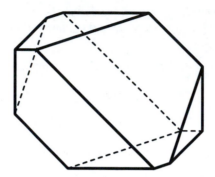

Figure 6.1 A resotope, when $d = 3$.

polytopes are quite varied. Figure 6.1 shows a resohedron (2-dimensional resotope) with five pentagonal faces and three triangular ones. To avoid these geometric difficulties we develop an approach, based on the Principle of Inclusion-Exclusion, that allows us to replace the complicated geometry of a general polytope with the simple geometry of a simplex.

The process of passing from the simplex of all accordions to an arbitrary resotope is complex. Indeed, as one chops corners off the simplex of all accordions to obtain a general resotope, interior accordions become boundary accordions, so their signatures change. Sometimes the removed simplices overlap, so one must restore any part that has been removed more than once. As we shall show, there is nonetheless a good way of handling it. Before we formulate this precisely, let us consider an example. The set of all Γ-resohedra with $d = 2$ and fixed value s is represented in Figure 6.2 by the triangle $\triangle\, \mathfrak{abc}$ with vertices

$$\mathfrak{a} = \left\{ \begin{matrix} s & & s & & 0 \\ & 0 & & s & \end{matrix} \right\}, \qquad \mathfrak{b} = \left\{ \begin{matrix} s & & 0 & & 0 \\ & s & & s & \end{matrix} \right\}, \qquad \mathfrak{c} = \left\{ \begin{matrix} s & & s & & s \\ & 0 & & 0 & \end{matrix} \right\}.$$

We are concerned with the shaded resotope $\mathcal{A} = \mathcal{A}_s^{\Gamma}(c_0, c_1, \infty)$, which is obtained by truncating the simplex $\triangle\, \mathfrak{abc}$ by removing $\triangle\, \mathfrak{aeg}$ and $\triangle\, \mathfrak{dbh}$. We use ∞ to mean any value of c_2 that is so large that the inequality $\mu_2 \leqslant c_2$ is automatically true (and strict) for all Γ-accordions; indeed any $c_2 > s$ can be replaced by ∞ without changing $\mathcal{A}$.

Then, since the $\triangle\, \mathfrak{def}$ has been removed twice, it must be restored, and we may write

$$\mathcal{A} = \triangle\, \mathfrak{abc} - \triangle\, \mathfrak{aeg} - \triangle\, \mathfrak{dbh} + \triangle\, \mathfrak{def}.$$

Now in this equation, $\triangle\, \mathfrak{def}$ (for example) should be regarded as an element of 3_Γ, and in addition to specifying its *support* – its underlying set – we must also specify which signatures occur with each accordion that appears in it, and with what sign. For example, in $\triangle\, \mathfrak{def}$ the accordion $\mathfrak{f}$ will occur with four different signatures: the actual contribution of $\mathfrak{f}$ to $\triangle\, \mathfrak{def}$ is

$$\mathfrak{f}_{\square\square*} - \mathfrak{f}_{\square**} - \mathfrak{f}_{*\square*} + \mathfrak{f}_{***} \,.$$

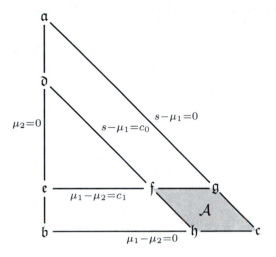

Figure 6.2 Inclusion-Exclusion.

Now let us give a formal description of this process. A signature η is called *nodal* if each η_i is either $\circ$ or $\square$. We fix a nodal signature η. Let $\mathrm{CP}_\eta(c_0, \cdots, c_d)$ be the "cut and paste" simplex, which is the set of Γ-accordions

$$\mathfrak{a} = \left\{ \begin{matrix} s & \mu_1 & \cdots & \mu_d \\ & \nu_1 & \cdots & \nu_d \end{matrix} \right\} \tag{6.21}$$

that satisfy the inequalities

$$\mu_i - \mu_{i+1} \geqslant c_i', \qquad c_i' = \begin{cases} c_i & \text{if } \sigma_i = \square, \\ 0 & \text{if } \sigma_i = \circ. \end{cases} \tag{6.22}$$

To see that this is truly a simplex, embed it into Euclidean space via the map

$$\mathfrak{a} \longmapsto (a_0, a_1, \cdots, a_d), \qquad a_i = \mu_i - \mu_{i+1} - c_i'.$$

The image of this map is the set of integer points in the simplex defined by the inequalities

$$a_i \geqslant 0, \qquad a_0 + \ldots + a_d \leqslant N, \qquad N = s - \sum c_i'.$$

Thus in the above example, with $\eta = \square\square\circ$ we have

$$\triangle\,\mathfrak{abc} = \mathrm{CP}_{\circ\circ\circ}(c_0, c_1, \infty), \qquad \triangle\,\mathfrak{aeg} = \mathrm{CP}_{\circ\square\circ}(c_0, c_1, \infty),$$

$$\triangle\,\mathfrak{dbh} = \mathrm{CP}_{\square\circ\circ}(c_0, c_1, \infty), \qquad \triangle\,\mathfrak{def} = \mathrm{CP}_{\square\square\circ}(c_0, c_1, \infty).$$

If σ and τ are signatures, we say that τ is a *subsignature* of σ if $\tau_i = \sigma_i$ whenever $\tau_i \neq *$, and we write $\tau \subset \sigma$ in this case. In other words, $\tau \subset \sigma$ if τ is obtained from σ by changing some $\square$'s or $\circ$'s to $*$'s. If τ is a signature, we will denote $\mathrm{sgn}(\tau) = (-1)^\varepsilon$ where ε is the number of boxes in τ.

Returning to the general case, let $\mathfrak{a}$ be a Γ-accordion, and let σ be a compatible signature for $\mathfrak{a}$. Define

$$\Lambda_\Gamma(\mathfrak{a}, \sigma) = \sum_{\mathfrak{a}\text{-compatible } \tau \subset \sigma} \mathrm{sgn}(\tau) \mathcal{G}_\Gamma(\mathfrak{a}, \tau),$$

$$\Lambda_\Delta(\mathfrak{a}', \sigma) = \sum_{\mathfrak{a}\text{-compatible } \tau \subset \sigma} \mathrm{sgn}(\tau) \mathcal{G}_\Delta(\mathfrak{a}', \tau).$$

In the definition of subsignature we allow either $\square$'s or $\circ$'s to be changed to $*$'s (needed for later purposes). But in this summation, because τ and σ are both required to be compatible with the same accordion $\mathfrak{a}$, only $\square$'s are changed to $*$ between σ and τ in the τ that appear.

Thus in the above example

$$\Lambda_\Gamma(\mathfrak{f}, \square\square*) = \mathcal{G}_\Gamma(\mathfrak{f}_{\square\square*}) - \mathcal{G}_\Gamma(\mathfrak{f}_{\square**}) - \mathcal{G}_\Gamma(\mathfrak{f}_{*\square*}) + \mathcal{G}_\Gamma(\mathfrak{f}_{***}) \,.$$

Now let η be a nodal signature, and we may take $c_i = \infty$ if $\eta_i = \circ$. Let $\mathfrak{a} \in \mathrm{CP}_\eta(c_0, \cdots, c_d)$. Let $\sigma = \sigma(\mathfrak{a})$ be the subsignature of η obtained by changing η_i to $*$ when the inequality (6.22) is strict.

Statement E. *Assume that $n|s$. We have*

$$\sum_{\mathfrak{a} \in \mathrm{CP}_\eta(c_0, \cdots, c_d)} \Lambda_\Gamma(\mathfrak{a}, \sigma) = \sum_{\mathfrak{a} \in \mathrm{CP}_\eta(c_0, \cdots, c_d)} \Lambda_\Delta(\mathfrak{a}', \sigma). \qquad (6.23)$$

We reiterate that in this sum σ depends on $\mathfrak{a}$, and we have described the nature of the dependence above.

We have already noted that $n|s$ can be imposed in Statement D and now we impose it explicitly. We will show in Chapter 14 that Statement E implies Statement D by application of the Inclusion-Exclusion principle. We hope for the purpose of this outline of the proof, the above example will make that plausible. In that example, the four triangles $\triangle\, \mathfrak{abc}$, $\triangle\, \mathfrak{aeg}$, $\triangle\, \mathfrak{dbc}$, and $\triangle\, \mathfrak{def}$ are examples of cut and paste simplices.

We have already mentioned that in the context of Statements B, C, or D, it is empirically possible to partition the sum into a disjoint union of smaller units called *packets* such that the identity is true when summation is restricted to a packet. Yet it is also true that in those contexts, a general rule describing the packets is notoriously slippery to nail down. However, in the context of Statement E we are able to describe the packets explicitly. The d-dimensional simplex $\mathrm{CP}_\eta(c_0, \cdots, c_d)$ is partitioned into *facets*, which are subsimplices of lower dimension. Specifically, there are $\binom{d+1}{f+1}$ facets that are simplices of dimension f; we will call these f-*facets*. We will define the packets so that if $\mathfrak{a}$ lies on the interior of an f-facet, then the packet containing $\mathfrak{a}$ has $\binom{d+1}{f+1}$ elements, one chosen from the interior of each r-facet.

Let us make this precise. First we observe the following description of the $d + 1$ vertices $\mathfrak{a}_i$ of $\mathrm{CP}_\eta(c_0, \cdots, c_d)$. If $0 \leqslant i \leqslant d$ let $\mathfrak{a}_i$ be the accordion whose coordinates μ_i are determined by the equations

$$\mu_j - \mu_{j+1} = c'_j, \qquad \text{for all } 0 \leqslant j \leqslant d \text{ with } j \neq i.$$

A *closed f-facet* of $\mathrm{CP}_\eta(c_0, \cdots, c_d)$ will be the set of integer points in the closed convex hull of a subset of cardinality $f + 1$ of the set $\{\mathfrak{a}_0, \cdots, \mathfrak{a}_d\}$ of vertices.

The subset of this f-facet consisting of elements that are not in any f'-facet for $f' < f$ is called an *open f-facet*. Clearly every element of $\mathrm{CP}_\eta(c_0, \cdots, c_d)$ lies in a unique open facet.

We associate the facets with subsignatures σ of η; if σ is obtained by replacing η_i ($i \in S$) by $*$, where S is some subset of $\{0, 1, 2, \cdots, d\}$, then we will denote the set of integer points in the closed (resp. open) convex hull of $\mathfrak{a}_i$ ($i \in S$) by $\overline{\mathcal{S}_\sigma}$ (resp. $\mathcal{S}_\sigma$). The facet $\overline{\mathcal{S}_\sigma}$ is itself a simplex, of dimension f. Thus if σ is a signature with exactly $f + 1$ $*$'s, we call σ an f-*signature* or an f-*subsignature of η.

Now if σ and τ are f-subsignatures of η, then we will define a bijection $\phi_{\sigma,\tau}$: $\overline{\mathcal{S}_\sigma} \longrightarrow \overline{\mathcal{S}_\tau}$. It is the unique affine linear map that takes the vertices of $\overline{\mathcal{S}_\sigma}$ to the vertices of $\overline{\mathcal{S}_\tau}$ in order. This means that if

$$S = \{s_0, \cdots, s_f\}, \qquad 0 \leqslant s_0 < s_1 < \cdots < s_f \leqslant d$$

is the set of i such that $\sigma_i = *$, and similarly if

$$T = \{t_0, \cdots, t_f\}, \qquad 0 \leqslant t_0 < t_1 < \cdots < t_f \leqslant d$$

is the set of i such that $\tau_i = *$, then $\phi_{\sigma,\tau}$ takes $\mathfrak{a}_{s_i}$ to $\mathfrak{a}_{t_i}$, and this map on vertices is extended by affine linearity to a map on all of $\overline{\mathcal{S}_\sigma}$.

It is obvious from the definition that $\phi_{\sigma,\sigma}$ is the identity map on $\mathcal{S}_\sigma$ and that if σ, τ, θ are f-subsignatures of η then $\phi_{\tau,\theta} \circ \phi_{\sigma,\tau} = \phi_{\sigma,\theta}$. This means that we may define an equivalence relation on $\mathrm{CP}_\eta(c_0, \cdots, c_d)$ as follows. Let $\mathfrak{a}, \mathfrak{b} \in \mathrm{CP}_\eta(c_0, \cdots, c_d)$. Let $\mathcal{S}_\sigma$ and $\mathcal{S}_\tau$ be the (unique) open facets such that $\mathfrak{a} \in \mathcal{S}_\sigma$ and $\mathfrak{b} \in \mathcal{S}_\tau$. Then $\mathfrak{a}$ is equivalent to $\mathfrak{b}$ if and only if $\phi_{\sigma,\tau}(\mathfrak{a}) = \mathfrak{b}$. The equivalence classes are called *packets*. It is clear from the definitions that the number $f + 1$ of $*$'s in σ is constant for σ that appear in a packet Π, and we will call Π a f-*packet*. Clearly every f-packet contains exactly one element from each f-simplex.

Statement F. *Assume that $n|s$. Let Π be a packet. Then*

$$\sum_{\mathfrak{a} \in \Pi} \Lambda_\Gamma(\mathfrak{a}, \sigma) = \sum_{\mathfrak{a} \in \Pi} \Lambda_\Delta(\mathfrak{a}', \sigma). \tag{6.24}$$

As in Statement E, σ depends on $\mathfrak{a}$ in this sum, and from the definition of packets, no σ appears more than once; in fact, if Π is an f-packet, then every f-subsignature σ of η appears exactly once on each side of the equation.

It is obvious that Statement F implies Statement E. There is one further sufficient condition that we call **Statement G**, but a proper formulation requires more notation than we want to give at this point. We will therefore postpone the formulation of Statement G to the end of Chapter 15, describing it here in informal terms.

Due to the knowability property of the products of Gauss sums that make up $\mathcal{G}_\Gamma(\mathfrak{a}, \sigma)$, these can be evaluated explicitly when $n|s$ (Proposition 15.1) and this leads to an evaluation of $\Lambda_\Gamma(\mathfrak{a}, \sigma)$ and a similar evaluation of $\Lambda_\Delta(\mathfrak{a}', \sigma)$ (Theorems 15.3 and 15.4). However, these evaluations depend on the divisibility properties of the μ_i that appear in the top row of $\mathfrak{a}$ given by (6.21); more precisely, if Σ is a subset of $\{1, 2, \cdots, d\}$, let $\delta_n(\Sigma; \mathfrak{a})$ be 1 if $n|\mu_i$ for all $i \in \Sigma$ and 0 otherwise. Then there is a sum over certain such subsets Σ of $\mathfrak{a}$ – only those i such that $\sigma_i \neq \bigcirc$ can appear – and the terms that appear are with a coefficient $\delta_n(\Sigma; \mathfrak{a})$. We recall that each σ appears only once on each side of (6.24), and hence $\mathfrak{a}$ is really a function of σ. Thus we are reduced to proving Statement G, amounting to an identity (15.11) in which there is first a sum over all f-subsignatures σ of η, and then a sum over subsets Σ, of $\{1, 2, \cdots, d\}$.

The identity (15.11) seems at first perplexing since $\delta_n(\Sigma; \mathfrak{a})$ depends on $\mathfrak{a}$. It won't work to simply interchange the order of summation since then $\delta_n(\Sigma; \mathfrak{a})$ will not be constant on the inner sum over $\mathfrak{a}$ (or equivalently σ). However we are able to identify an equivalence relation that we call *concurrence* on pairs (σ, Σ) such that $\delta_n(\Sigma; \mathfrak{a})$ is constant on concurrence classes (Proposition 16.2). We will then need a result that implies that some groups of terms from the same side of (15.11) involve concurrent data (Proposition 16.5). These concurrent data are called Γ-*packs* for the left-hand side or Δ-*packs* for the right-hand side. Then we will need a rather more subtle result (Proposition 16.10) giving a bijection between the Γ-packs and the Δ-packs that also matches concurrent data. With these combinatorial preparations, we will be able to prove (15.11) and therefore Statement G in Chapter 17.

Chapter Seven

Statement B Implies Statement A

In this chapter we will recall the use of the Schützenberger involution on Gelfand-Tsetlin patterns in [15] to prove that Statement B implies Statement A. We will return to these statements two more times in the later chapters of the book. In Chapter 18 we will reinterpret both Statements A and B in terms of crystals, and directly prove that the reinterpreted Statement B implies the reinterpreted Statement A in Theorem 18.2. Then in Chapter 19 we will yet again reinterpret Statements A and B in a different context, and yet again directly prove that the reinterpreted Statement B implies the reinterpreted Statement A in Theorem 19.10.

We observe that the Schützenberger involution q_r can be formulated in terms of operations on short Gelfand-Tsetlin patterns. If

$$
\mathfrak{T} = \left\{
\begin{array}{ccccccc}
a_{00} & & a_{01} & & a_{02} & \cdots & a_{0r} \\
& a_{11} & & a_{12} & & & a_{1r} \\
& & \ddots & & & \reflectbox{$\ddots$} & \\
& & & a_{rr} & & &
\end{array}
\right\}
$$

is a Gelfand-Tsetlin pattern and $1 \leqslant k \leqslant r$, then extracting the $r - k$, $r + 1 - k$ and $r + 2 - k$ rows gives a short Gelfand-Tsetlin pattern t. Replacing this with the pattern t′ gives a new Gelfand-Tsetlin pattern, which is the one denoted $t_r \mathfrak{T}$ in Chapter 1. Thus

$$
t_1 \left\{
\begin{array}{ccccc}
\lambda_1 & & \lambda_2 & & \lambda_3 \\
& a & & b & \\
& & c & &
\end{array}
\right\}
=
\left\{
\begin{array}{ccccc}
\lambda_1 & & \lambda_2 & & \lambda_3 \\
& a & & b & \\
& & a + b - c & &
\end{array}
\right\}
$$

and

$$
t_2 \left\{
\begin{array}{ccccc}
\lambda_1 & & \lambda_2 & & \lambda_3 \\
& a & & b & \\
& & c & &
\end{array}
\right\}
=
\left\{
\begin{array}{ccccc}
\lambda_1 & & \lambda_2 & & \lambda_3 \\
& a' & & b' & \\
& & c & &
\end{array}
\right\}
$$

where $a' = \lambda_1 + \max(\lambda_2, c) - a$ and $b' = \lambda_3 + \min(\lambda_2, c) - b$.

We defined q_0 to be the identity map, and defined $q_i = t_1 t_2 \cdots t_i q_{i-1}$. The t_i have order two. They do not satisfy the braid relation, so $t_i t_{i+1} t_i \neq t_{i+1} t_i t_{i+1}$. However $t_i t_j = t_j t_i$ if $|i - j| > 1$ and this implies that the q_i also have order two. We note that

$$
q_i = q_{i-1} q_{i-2} t_i q_{i-1}. \tag{7.1}
$$

Let $A_i = \sum_j a_{i,j}$ be the sum of the i-th row of $\mathfrak{T}$. It may be checked that the row sums of $q_r \mathfrak{T}$ are (in order)

$$
A_0, A_0 - A_r, A_0 - A_{r-1}, \cdots, A_0 - A_1.
$$

From this it follows that

$$k_\Gamma(q_r \mathfrak{T}) = k_\Delta(\mathfrak{T}).$$

From this we see that Statement A will follow if we prove

$$\sum_{k_\Gamma(\mathfrak{T})=k} G_\Gamma(\mathfrak{T}) = \sum_{k_\Gamma(\mathfrak{T})=k} G_\Delta(q_r \mathfrak{T}). \tag{7.2}$$

We note that the sums are over all patterns with fixed *top row* and *row sums*.

Let us denote

$$G_R^i(\mathfrak{T}) = \prod_{j=i}^{r} \begin{cases} g(\Gamma_{ij}) & \text{if } \Gamma_{ij} \text{ is boxed but not circled in } \Gamma(\mathfrak{T}); \\ q^{\Gamma_{ij}} & \text{if } \Gamma_{ij} \text{ is circled but not boxed;} \\ h(\Gamma_{ij}) & \text{if } \Gamma_{ij} \text{ neither circled nor boxed;} \\ 0 & \text{if } \Gamma_{ij} \text{ both circled and boxed} \end{cases}$$

and

$$G_L^i(\mathfrak{T}) = \prod_{j=i}^{r} \begin{cases} g(\Delta_{ij}) & \text{if } \Delta_{ij} \text{ is boxed but not circled in } \Gamma(\mathfrak{T}); \\ q^{\Delta_{ij}} & \text{if } \Delta_{ij} \text{ is circled but not boxed;} \\ h(\Delta_{ij}) & \text{if } \Delta_{ij} \text{ neither circled nor boxed;} \\ 0 & \text{if } \Delta_{ij} \text{ both circled and boxed,} \end{cases}$$

where Γ_{ij} and Δ_{ij} are given by (1.11). Thus

$$G_\Gamma(\mathfrak{T}) = \prod_{i=1}^{r} G_R^i(\mathfrak{T}), \qquad G_\Delta(\mathfrak{T}) = \prod_{i=1}^{r} G_L^i(\mathfrak{T}).$$

To facilitate our inductive proof, if $i \leqslant r$ let $\mathfrak{T}_i$ denote the pattern formed with the bottom $i+1$ rows of $\mathfrak{T}$. Let $^{(i)}G_\Gamma(\mathfrak{T}) = G_\Gamma(\mathfrak{T}_i)$ and $^{(i)}G_\Delta(\mathfrak{T}) = G_\Delta(\mathfrak{T}_i)$. Then

$$\begin{aligned}
\sum_{k_\Gamma=k} G_\Gamma(\mathfrak{T}) &= \sum_{k_\Gamma=k} G_R^r(\mathfrak{T}) \cdot {}^{(r-1)}G_\Gamma(\mathfrak{T}) \\
&= \sum_{k_\Gamma=k} G_R^r(\mathfrak{T}) \cdot {}^{(r-1)}G_\Delta(q_{r-1}\mathfrak{T}) \\
&= \sum_{k_\Gamma=k} G_R^r(\mathfrak{T}) \cdot G_L^{r-1}(q_{r-1}\mathfrak{T}) \cdot {}^{(r-2)}G_\Delta(q_{r-1}\mathfrak{T}) \\
&= \sum_{k_\Gamma=k} G_R^r(\mathfrak{T}) \cdot G_L^{r-1}(q_{r-1}\mathfrak{T}) \cdot {}^{(r-2)}G_\Gamma(q_{r-2}q_{r-1}\mathfrak{T}) \\
&= \sum_{k_\Gamma=k} G_R^r(q_{r-2}q_{r-1}\mathfrak{T}) \cdot G_L^{r-1}(q_{r-2}q_{r-1}\mathfrak{T}) \cdot {}^{(r-2)}G_\Gamma(q_{r-2}q_{r-1}\mathfrak{T}).
\end{aligned}$$

Here the first step is by definition; the second step is by applying the induction hypothesis that Statement A is true for $r-1$ to $\mathfrak{T}_{r-1}$; the third step is by definition; the fourth step is by induction, using Statement A for $r-2$ applied to $\mathfrak{T}_{r-2}$; and the last step is because $q_{r-2}q_{r-1}$ does not change the top two rows of $\mathfrak{T}$, hence does not affect the value of G_R^r, and similarly q_{r-2} does not change the value of G_R^{r-1}.

On the other hand we have

$$\sum_{k_\Gamma=k} G_\Delta(q_r\mathfrak{T}) = \sum_{k_\Gamma=k} G_L^r(q_r\mathfrak{T}) \cdot {}^{(r-1)}G_\Delta(q_r\mathfrak{T})$$

$$= \sum_{k_\Gamma=k} G_L^r(q_r\mathfrak{T}) \cdot {}^{(r-1)}G_\Gamma(q_{r-1}q_r\mathfrak{T})$$

$$= \sum_{k_\Gamma=k} G_L^r(q_r\mathfrak{T}) \cdot G_R^{r-1}(q_{r-1}q_r\mathfrak{T}) \cdot {}^{(r-2)}G_\Gamma(q_{r-1}q_r\mathfrak{T})$$

$$= \sum_{k_\Gamma=k} G_L^r(q_{r-1}q_r\mathfrak{T}) \cdot G_R^{r-1}(q_{r-1}q_r\mathfrak{T}) \cdot {}^{(r-2)}G_\Gamma(t_r q_{r-1}q_r\mathfrak{T}).$$

Here the first step is by definition, the second by induction, the third by definition, and the fourth because q_{r-1} does not affect the top two rows of $q_{r-1}q_r\mathfrak{T}$, and t_r does not affect the rows of $(q_{r-1}q_r\mathfrak{T})_{r-2}$. Now we use the assumption that Statement B is true. Statement B implies that

$$\sum G_L^r(q_{r-1}q_r\mathfrak{T}) \cdot G_R^{r-1}(q_{r-1}q_r\mathfrak{T}) = \sum G_R^r(t_r q_{r-1}q_r\mathfrak{T}) \cdot G_L^{r-1}(t_r q_{r-1}q_r\mathfrak{T})$$

where in this summation we may collect together all $q_{r-1}q_r\mathfrak{T}$ with the same first, third, fourth, ... rows and let only the second row vary to form a summation over short Gelfand-Tsetlin pattern. Substituting this back into the last identity gives

$$\sum_{k_\Gamma=k} G_\Delta(q_r\mathfrak{T}) = \sum_{k_\Gamma=k} G_R^r(t_r q_{r-1}q_r\mathfrak{T}) \cdot G_L^{r-1}(t_r q_{r-1}q_r\mathfrak{T}) \cdot {}^{(r-2)}G_\Gamma(t_r q_{r-1}q_r\mathfrak{T}).$$

Now we make use of (7.1) in the form $t_r q_{r-1}q_r = q_{r-2}q_{r-1}$ to complete the proof of Statement A, assuming Statement B.

Chapter Eight

Cartoons

This chapter will introduce a method of marking up a short Gelfand-Tsetlin pattern based on inequalities between its entries, that encodes the effect of the involution $t \mapsto t'$ and the boxing and circling of its accordion. This will have another benefit: it will lead to the decomposition of the pattern into pieces called *episodes* that will ultimately lead to the reduction to the totally resonant case.

PROPOSITION 8.1 *(i) If $n \nmid a$ then $h(a) = 0$ and $|g(a)| = q^{a-\frac{1}{2}}$.*
(ii) If $n|a$ then

$$h(a) = \phi(p^a) = q^{a-1}(q-1), \qquad g(a) = -q^{a-1}$$

(iii) If $n|a$ and $b > 0$ then

$$h(a+b) = q^a h(b), \qquad g(a+b) = q^a g(b).$$

(iv) If $n \nmid a, b$ but $n|a+b$ then

$$g(a)g(b) = q^{a+b-1}.$$

Property (iii) means that g^b and h^b are periodic with period n.

Proof. This is easily checked using standard properties of Gauss sums. □

To define the cartoon, we will take a slightly more formal approach to the short Gelfand-Tsetlin patterns. Let

$$\Theta = \{(i,j) \in \mathbb{Z} \times \mathbb{Z} | 0 \leqslant i \leqslant 2, 0 \leqslant j \leqslant d+1-i\}.$$

We call this set the *substrate*, and divide Θ into three *rows*, which are

$$\Theta_0 = \{(0,j) \in \Theta | 0 \leqslant j \leqslant d+1\},$$
$$\Theta_1 = \{(1,j) \in \Theta | 0 \leqslant j \leqslant d\},$$
$$\Theta_2 = \{(2,j) \in \Theta | 0 \leqslant j \leqslant d-1\},$$

Let $\Theta_B = \Theta_1 \cup \Theta_2$. Each row has an order in which $(i,j) \leqslant (i,j')$ if and only if $j \leqslant j'$.

Now we can redefine a *short Gelfand-Tsetlin pattern* to be an integer valued function t on the substrate, subject to the conditions that we have already stated. Thus the short Gelfand-Tsetlin pattern (6.2) corresponds to the function on Θ such that $l_i = t(0,i)$, $a_i = t(1,i)$ and $b_i = t(2,i)$. The Γ and Δ preaccordions, which are arrays having only two rows, may similarly be described as functions on Θ_B (the bottom and middle rows) in the same way. Specifying the circled and boxed elements just means specifying subsets of Θ_B.

Now the vertices of the cartoon will be the elements of the substrate Θ, and we have only to define the edges. With t as in (6.2) we connect $(1, i)$ to $(0, i)$ if either $i = 0$ or $l_i \leqslant b_{i-1}$, and we connect $(1, i)$ to $(2, i - 1)$ if $i > 0$ and $b_{i-1} \leqslant l_i$. Furthermore we connect $(1, i)$ to $(0, i + 1)$ if either $i = d$ or if $i < d$ and $l_{i+1} \geqslant b_i$, and we connect $(1, i)$ to $(2, i)$ if $i < d$ and $b_i \geqslant l_{i+1}$. For example, consider the short pattern of rank 5:

$$
t = \left\{
\begin{array}{ccccccc}
23 & & 15 & & 12 & 5 & & 2 & & 0 \\
& 20 & & 12 & & 5 & & 4 & & 2 \\
& 14 & & 9 & & 5 & & 3 &
\end{array}
\right\}
\tag{8.1}
$$

It is convenient to draw the cartoon as a graph on top of the preaccordion representing t, as follows:

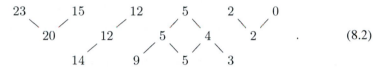

$$\tag{8.2}$$

- The cartoon depends only on the top and bottom rows of t, so it is really a function of the prototype $\mathfrak{S}$ to which t belongs.

- The cartoon encodes the relationship between t and t'. Indeed, suppose that the cartoon has a subgraph $x \,\text{---}\, y \,\text{---}\, z$ where y is in the middle row, x and z are each in either the top or bottom row, with x is to the left of y and z is to the right. Then in t', y is replaced by $x + z - y$.

For example, if t is given by (8.1), then the cartoon (8.2) tells us how to compute

$$
t' = \left\{
\begin{array}{ccccccc}
23 & & 15 & & 12 & 5 & & 2 & & 0 \\
& 18 & & 14 & & 9 & & 4 & & 0 \\
& 14 & & 9 & & 5 & & 3 &
\end{array}
\right\} ;
$$

the middle row entries are $18 = 23 + 15 - 20$, $14 = 12 + 14 - 12$, $9 = 5 + 9 - 5$, $4 = 5 + 3 - 4$ and $0 = 2 + 0 - 2$.

The connected components of the cartoon are called *episodes*. These may be arranged in an order $\mathcal{E}_1, \cdots, \mathcal{E}_N$ so that if $i < j$ and $\alpha \in \mathcal{E}_i$, $\beta \in \mathcal{E}_j$, and if α and β are in the same row of the substrate Θ, then $\alpha < \beta$. With this partial order, if $\alpha \in \mathcal{E}_i$, $\beta \in \mathcal{E}_j$ then $t(\alpha) > t(\beta)$ regardless of whether or not α and β are in the same row.

We call short pattern (6.2) *resonant* at i if $l_{i+1} = b_i$. A *resonance* of order k is a sequence $R = \{i, i + 1, \cdots, i + k - 1\}$ such that t is resonant at each $j \in R$; the sequence must be maximal with this property, so that t is not resonant at $i - 1$ or $i + k$. In the example (8.1), a resonance at 2 can be recognized from the cartoon by the diamond shape between $l_3 = b_2 = 5$.

We will next describe another kind of diagram related to the cartoon in which we mark certain edges with double bonds, and box and circle certain vertices. We will refer to the diagram in which the bonded edges and circled vertices are marked as the *bond-marked* cartoon. See (8.3) and (8.4) below for examples.

- Unlike the cartoon, the bond-marked cartoon really depends on t, not just on its prototype.

- The bond-marked cartoon is useful since the circling and boxing of the Γ and Δ preaccordions can be read off from it.

The edge joining $\alpha, \beta \in \Theta$ will be called *distinguished* if $t(\alpha) = t(\beta)$. In representing the bond-marked cartoon graphically we will mark the distinguished edges by double bonds, which may be read as equal signs. Thus in the example (8.1), the cartoon of t becomes

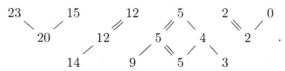

(We have drawn this labeling on top of t itself, but ultimately we will draw it on top of the Γ or Δ preaccordions.)

We observe that while the original bond-unmarked cartoon only depends on the pattern prototype $\mathfrak{S}$ to which t belongs, this diagram does depend on t. In particular, t and t′ no longer have the same cartoon, since the double bonds move under the involution $t \longmapsto t'$. However the rule is quite simple:

LEMMA 8.2 *Suppose the bond-marked cartoon of* t *has a subgraph of the form* $x - z = z$, *where the first z is in the middle row, so that x and the second z are in the top or bottom row. Then in the bond-marked cartoon the double bond moves to the other edge, so the bond-marked cartoon of* t′ *contains a subgraph* $x = x - z$.

Proof. Immediate from the definitions.   □

In this example, the bond-marked cartoon of t′ is

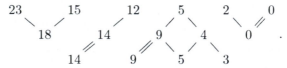

As we have already mentioned, the cartoon is very useful when superimposed on the Γ and Δ' preaccordions, where $\Gamma = \Gamma_t$ and $\Delta' = \Delta_{t'}$. Since these arrays have only two rows, we add a third row at the top. We will also box and circle certain entries, by a convention that we will explain after giving an example. Thus in this example

$$\Gamma = \left\{ \begin{matrix} 9 & & 4 & & 4 & & 4 & & 2 \\ & 6 & & 9 & & 9 & & 10 & \end{matrix} \right\}, \qquad \Delta' = \left\{ \begin{matrix} 5 & & 6 & & 9 & & 10 & & 12 \\ & 4 & & 4 & & 4 & & 3 & \end{matrix} \right\}.$$

We superimpose the cartoon on these, representing Γ thus:

and Δ' as

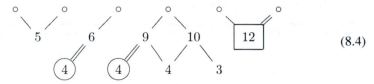

$$(8.4)$$

We've inserted a row of ∘'s in the top (0-th) row since the Γ and Δ' preaccordions have first and second rows but no 0-th row; we supply these for the purpose of drawing the bond-marked cartoon. When the bond-marked cartoon is thus placed on top of the Γ and Δ' preaccordions, the circling and boxing conventions can be conveniently understood.

- In the first row of Γ or the second row of Δ' we circle an entry if a double bond is above it and to the right. We box an entry if a double bond is above it and to the left. Thus:

- In the second row of Γ or the first row of Δ' we circle an entry if a double bond is above it and to the left. We box an entry if a double bond is above it and to the right. Thus:

Now we have the basic language that will allow us to prove the reduction to the totally resonant case.

Chapter Nine

Snakes

The key lemma of this chapter was stated without proof in [15], where it was called the "Snake Lemma." (It is of course not the famous Snake Lemma from homological algebra.) Here we will recall it, prove it, and use it to prove the statement made in Chapter 6, that (6.13) is "often" true.

By an *indexing* of the Γ preaccordion we mean a bijection

$$\phi : \{1, 2, \cdots, 2d + 1\} \longrightarrow \Theta_B.$$

With such an indexing in hand, we will denote $\Gamma_t(\alpha)$ by $\gamma_k(t)$ or just γ_k if $\alpha = \phi(k)$ corresponds to k. Thus

$$\{\gamma_1, \gamma_2, \cdots, \gamma_{2d+1}\} = \{\Gamma(\alpha) \,|\, \alpha \in \Theta_B\}.$$

We will also consider an indexing ψ of the Δ' preaccordion, and we will denote $\Delta'(\alpha)$ by δ_k' if $\alpha = \psi(k)$. It will be convenient to extend the indexings by letting $\gamma_0 = \gamma_{2d+2} = 0$ and $\delta_0' = \delta_{2d+2}' = 0$.

PROPOSITION 9.1 *There exist indexings of the Γ and Δ' preaccordions such that*

$$\delta_k' = \begin{cases} \gamma_k & \text{if } k \text{ is even,} \\ \gamma_k + \gamma_{k-1} - \gamma_{k+1} & \text{if } k \text{ is odd.} \end{cases} \tag{9.1}$$

If $i \in \{1, 2, \cdots, 2d + 2\}$, and if $\phi(i) \in \mathcal{E}_k$, then $\psi(i) \in \mathcal{E}_k$ also. Moreover if $\phi(j), \psi(j) \in \mathcal{E}_l$ and $k < l$ then $i < j$.

Before we prove this, let us confirm it in the specific example at hand. With Γ and Δ' as in (8.3) and (8.4), we may take the correspondence as follows:

k	1	2	3	4	5	6	7	8	9
(i, j) in Γ order	$(1, 0)$	$(1, 1)$	$(2, 0)$	$(1, 2)$	$(2, 1)$	$(1, 3)$	$(2, 2)$	$(2, 3)$	$(1, 4)$
episode	$\mathcal{E}_1$	$\mathcal{E}_2$	$\mathcal{E}_2$	$\mathcal{E}_3$	$\mathcal{E}_3$	$\mathcal{E}_3$	$\mathcal{E}_3$	$\mathcal{E}_3$	$\mathcal{E}_4$
γ_k	9	4	6	4	9	4	9	10	2
(i, j) in Δ' order	$(1, 0)$	$(2, 0)$	$(1, 1)$	$(2, 1)$	$(1, 2)$	$(2, 2)$	$(2, 3)$	$(1, 3)$	$(1, 4)$
δ_k'	5	4	6	4	9	4	3	10	12
episode	$\mathcal{E}_1$	$\mathcal{E}_2$	$\mathcal{E}_2$	$\mathcal{E}_3$	$\mathcal{E}_3$	$\mathcal{E}_3$	$\mathcal{E}_3$	$\mathcal{E}_3$	$\mathcal{E}_4$

$$\tag{9.2}$$

The meaning of the last assertion of Proposition 9.1 is that each indexing visits the episodes of the sequence in order from left to right, and after it is finished with an episode, it moves on to the next with no skipping around. Thus both indexings must be in the same episode.

The reason that Proposition 9.1 was called the "Snake Lemma" in [15] is that if one connects the nodes of Θ_B in the indicated orderings, a pair of "snakes" becomes visible. Thus in the example (9.2) the paths will look as follows. The Γ indexing is represented:

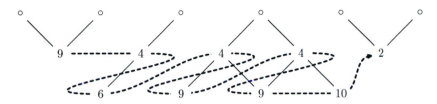

and the Δ' indexing:

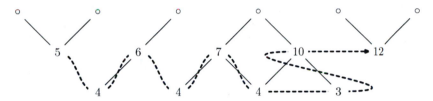

- The proof will provide a particular description of the pair of snakes; in applying Proposition 9.1 will sometimes need this particular description. We will describe the pair of indexings, or "snakes," as *canonical* if they are produced by the method described in the proof, which is expressed in Table 9.1. Thus we implicitly use the proof as well as the statement of the Lemma.

- If there are resonances, there will be more than one possible pair of snakes. (Indeed, the reader will find another way of drawing the snakes in the preceding example.) These will be obtained through a process of *specialization* that will be described in the proof. Any one of these pairs of snakes will be described as canonical.

Proof. For this proof, double bonds are irrelevant, and we will work with the bond-unmarked cartoon. Thus both Γ and Δ' are again represented by the same cartoon, which in the example (8.1) was the cartoon (8.2). Resonances are a minor complication, which we eliminate as follows. We divide the cartoon into *panels*, each being of one five types:

t	T	B	b	R

The first panel of the cartoon is always of type t and the last one of type b. We call a cartoon *simple* if it contains no panels of type R.

The panel type R occurs at each resonance. Including it in our discussion would unnecessarily increase the number of cases to be considered, so we resolve each resonance by arbitrarily replacing each R by either a T or B. This will produce a simple cartoon. We refer to this process as *specialization*.

For example the cartoon (8.2) corresponds to the word $tBBRTb$, meaning that these panels appear in sequence from left to right. We replace the resonant panel R arbitrarily by either T or B; for example if we choose B we obtain the simple cartoon:

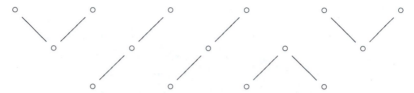

From the simple cartoon we may describe the algorithm for finding the pair of snakes, that is, the Γ and Δ' indexings. Each connected component (episode) in the simple cartoon has three vertices, the middle one being in the second row. We may classify these episodes into four classes as follows:

Class I	Class II	Class III	Class IV
○ ○	○ ○	○ ○	○ ○
○	○	○	○
○ ○	○ ○	○ ○	○ ○

Now we can describe the snakes. For each episode, we select a path from Table 9.1. The nodes labeled $\star$ will turn out to be indexed by even integers, and the nodes labeled $\bullet$ will be indexed by odd integers. We've subscripted the $\star$'s to indicate which entries are corresponding in the Γ and Δ'. A ? means that the information at hand does not determine whether the entry will be even or odd in the indexing, so we do not attempt to assign it a $\star$ or $\bullet$.

There are modifications at the left and right edges of the pattern: for example, if the first two panels are tB then the first connected component is of Class II, and the left parts of the Γ and Δ' indexings indicated in the table are missing. Thus we have a modification of the Class II pattern that we call II_t.

	Γ indexing	Δ' indexing
Class II_t		

Similarly, there are Classes III_t, II_b, and IV_b that can occur at the left or right edge of the pattern. In every case these are obtained by simply deleting part of the corresponding pattern, and we will not enumerate these for this reason.

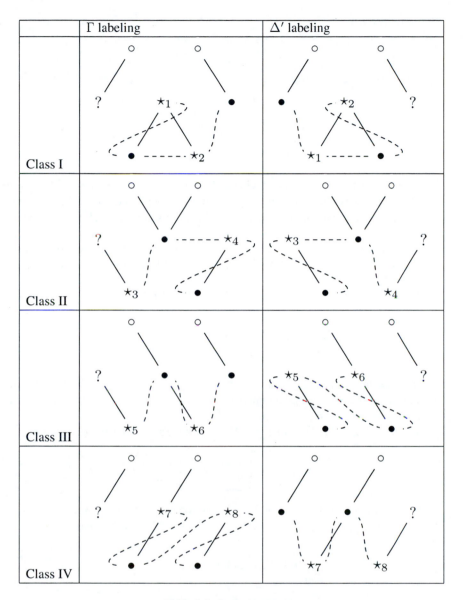

Table 9.1 Snake taxonomy.

It is necessary to see that these paths are assigned consistently. For example, suppose that the cartoon contains consecutive panels TBT. Inside the resulting configuration are both a Class II connected component and a Class I component. Referring to Table 9.1, both these configurations mandate the following dashed line in the Γ diagram.

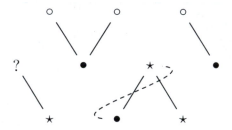

This sort of consistency must be checked in eight cases, which we leave to the reader. Then it is clear that splicing together the segments prescribed this way gives a consistent pair of snakes, and the indexings can be read off from left to right.

It remains, however, for us to prove (9.1). This is accomplished by four Lemmas. There are many things to verify; we will do one and leave the rest to the reader.

LEMMA 9.2 *If the j-th connected component is of Class I, then*

$$\Delta'(1, j) = \Gamma(2, j)$$
$$\Delta'(2, j - 1) = \Gamma(1, j)$$
$$\Delta'(2, j) = \Gamma(2, j - 1) + \Gamma(1, j) - \Gamma(2, j).$$

This asserts a part of (9.1), namely the equality $\delta'_k = \gamma_k$ for the vertices labeled $\star_1$ and $\star_2$ in Table 9.1, and the equality $\delta'_k = \gamma_k + \gamma_{k-1} - \gamma_{k+1}$ for the unstarred vertex in the connected component.

We prove that $\Delta'(1, j) = \Gamma(2, j)$. With a'_i defined by (6.5) and (6.6) we have

$$\Delta'(1, j) = \left(\sum_{i \leqslant j} l_i\right) - \left(\sum_{i \leqslant j} a'_i\right).$$

Our assumption that the j-th connected component is of Class I means that $l_j \geqslant b_{j-1}$ and that $l_{j+1} \geqslant b_j$, so $a'_j = b_{j-1} + b_j - a_j$. Moreover, since $l_j \geqslant b_{j-1}$ we have

$$\sum_{i < j} [\min(l_i, b_{i-1}) + \max(l_{i+1}, b_i)] =$$

$$\max(l_j, b_{j-1}) + \sum_{i < j} [\min(l_i, b_{i-1}) + \max(l_i, b_{i-1})] =$$

$$l_j + \sum_{i < j} (l_i + b_{i-1}) = \sum_{i \leqslant j} l_i + \sum_{i \leqslant j-2} b_i.$$

By (6.5) we therefore have

$$\sum_{i \leqslant j} a'_i = \sum_{i \leqslant j} l_i + b_i - a_i, \qquad \Delta'(1, j) = \sum_{i \leqslant j} a_i - b_i = \Gamma(2, j).$$

We leave the remaining two statements to the reader.

LEMMA 9.3 *If the j-th connected component is of Class II, then*

$$\Delta'(1,j) = \Gamma(1,j) + \Gamma(2, j-1) - \Gamma(1, j+1).$$

LEMMA 9.4 *If the j-th connected component is of Class III, then*

$$\Delta'(1,j) = \Gamma(2,j),$$
$$\Delta'(2,j) = \Gamma(1,j) + \Gamma(2, j-1) - \Gamma(2,j).$$

LEMMA 9.5 *If the j-th connected component is of Class IV, then*

$$\Delta'(2, j-1) = \Gamma(1,j),$$
$$\Delta'(1,j) = \Gamma(2, j-1) + \Gamma(1,j) - \Gamma(1, j+1).$$

We leave the proofs of the last three Lemmas to the reader. The assertions in (9.1) are contained in the Lemmas, for each δ'_k whose corresponding vertex is in the given connected component. It must lie in the first (middle) or second row of the cartoon, which is why there are three identities for Class I, one for Class II, and two for Classes III and IV. Thus (9.1) is proved for every δ'_k. The final assertion, that the episodes of the cartoon are visited from left to right in order by both indexings, can be seen by inspection from Table 9.1. □

LEMMA 9.6 (Circling Lemma) *Assume that t is strict.*
(i) Suppose that either of the following two configurations occurs in either Γ_t or Δ_t:

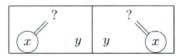

Then $x = y$.
(ii) If x occurs circled in either the Γ or Δ preaccordions of a strict pattern t, then either the same value x also occurs uncircled (and unboxed) at another location, or $x = 0$.

Proof. The first statement follows from the definition. To prove the second statement, we note that $y = x$ is unboxed since the pattern is strict. If it is uncircled, (ii) is proved. If it is circled, we continue to the right (if y is to the right of x) or to the left (if y is to the left of x) until we come to an uncircled one. This can only fail if we come to the edge of the pattern. If this happens, then $x = 0$. □

Chapter Ten

Noncritical Resonances

We recall that a short pattern (6.2) is *resonant* at i if $l_{i+1} = b_i$. This property depends only on the associated prototype, so resonance is actually a property of prototypes. We also call a first (middle) row entry a_i *critical* if it is equal to one of its four *neighbors*, which are l_i, l_{i+1}, b_i and b_{i-1}. We say that the resonance at i is *critical* if either a_i or a_{i+1} is critical.

THEOREM 10.1 *Suppose that* t *is a strict pattern with no critical resonances; then* t' *is also strict with no critical resonances. Choose canonical indexings* γ_i *and* δ'_i *as in Proposition 9.1. Then either* $G_\Gamma(t) = G_\Delta(t') = 0$ *or* $n|\gamma_i$*. In any case, we have*

$$G_\Gamma(t) = G_\Delta(t').$$

As an example, the pattern is called *superstrict* if the inequalities (6.7) and (6.8) are strict, that is, if

$$\min(l_j, b_{j-1}) > a_j > \max(l_{j+1}, b_j), \qquad 0 < j < d, \tag{10.1}$$

$$l_1 > a_1 > \max(l_2, b_1), \qquad \min(l_d, b_{d-1}) > a_d > l_{d+1}. \tag{10.2}$$

Thus if the patterns within a type are regarded as lattice points in a polytope, the superstrict patterns are the interior points. Again, the pattern (or prototype) is called *nonresonant* if there are no resonances. The theorem is clearly applicable if t is either superstrict or nonresonant.

Proof. To see that t' is strict, let a_i, b_i, l_i, and a'_i be as in (6.2) and (6.4). If t' is not strict, we must have $a'_i = a'_{i-1}$ for some i, and it is easy to see that this implies that $l_i = b_{i-1}$, and that t has a critical resonance at i. It is also easy to see that if t' has a critical resonance at i so does t.

Choose a pair of canonical indexings of $\Gamma = \Gamma_t$ and $\Delta' = \Delta_{t'}$. Our first task is to show that either $G_\Gamma(t) = G_\Delta(t') = 0$ or $n|\gamma_i$ for all even i. It is easy to see that γ_i and δ'_i are not boxed, since if they were, examination of every case in Table 9.1 shows that it would be at the terminus of a double bond in the bond-marked cartoon that is not one of the marked bonds in the figures. This could conceivably happen since in the proof of Proposition 9.1 we began by replacing the cartoon by a simple cartoon, a process that can involve discarding some parallel pairs of the bonds; however it would force γ_i (or δ'_i) to be a neighbor of a critical resonance, and we are assuming that t has no critical resonances.

Suppose that γ_i is not circled (i even). Then $G_\Gamma(t)$ is a multiple of $h(\gamma_i)$, which vanishes unless $n|\gamma_i$. If γ_i is circled, we must argue differently. By the Circling

Lemma (Lemma 9.6), either the same value γ_i occurs uncircled and unboxed somewhere in the Γ preaccordion, in which case $G_\Gamma(t)$ is again a multiple of $h(\gamma_i)$, or $\gamma_i = 0$. Since $n|\gamma_i$ if $\gamma_i = 0$ the conclusion that $G_\Gamma(t) = 0$ or $n|\gamma_i$ is proved. Since $\gamma_i = \delta'_i$ when n is even, we may also conclude that $G_\Delta(t') = 0$ unless the γ_i (i even) are all divisible by n.

We assume for the remainder of the proof that $n|\gamma_i$ when i is even. Let us denote

$$\tilde{\gamma}_i = \begin{cases} q^{\gamma_i} & \text{if } \gamma_i \text{ is circled in the } \Gamma \text{ indexing,} \\ g(\gamma_i) & \text{if } \gamma_i \text{ is boxed in the } \Gamma \text{ indexing,} \\ h(\gamma_i) & \text{otherwise,} \end{cases}$$

with $\tilde{\delta}'_i$ defined similarly. Thus

$$G_\Gamma(t) = \prod \tilde{\gamma}_i, \qquad G_\Delta(t') = \prod \tilde{\delta}'_i.$$

We next show that

$$\prod_{i \text{ even}} \tilde{\gamma}_i = \prod_{i \text{ even}} \tilde{\delta}'_i. \tag{10.3}$$

Since $\gamma_i = \delta'_i$ when i is even, and since as we have noted these entries are never boxed, the only way this could fail is if one of $\tilde{\gamma}_i$ and $\tilde{\delta}'_i$ is circled and the other not. We look at the connected component in the bond-unmarked cartoon containing $\tilde{\gamma}_i$. In Table 9.1, this entry is starred and must correspond to one of $\star_1$, $\star_2$, $\star_6$, or $\star_7$. (Since the snake is obtained by splicing pieces together different pieces of Table 9.1 it may also appear in $\star_3$, $\star_4$, $\star_5$, or $\star_8$.) If it is $\star_6$, then it is circled in the Γ indexing if and only if the bond above it is doubled, and by Lemma 8.2 the bond above $\star_6$ in the Δ' indexing is also doubled, so $\star_6$ is circled in both indexings; and similarly with $\star_7$. Turning to the Class I components, it is impossible for $\star_1$ to be circled in the Γ indexing, since this would imply a critical resonance; and $\star_2$ is never circled in the Δ' indexing for the same reasoning. Nevertheless it is possible for $\star_2$ to be circled in the Γ indexing but not the Δ' indexing. In this case, Lemma 8.2 shows that $\star_2$ is starred in the Δ' indexing but not the Γ indexing. This happens when the labeling of a Class I component looks like:

Thus if $\star_1$ is the i-th vertex in both orderings we have

$$\tilde{\gamma}_i = h(\gamma_i), \qquad \tilde{\gamma}_{i+2} = q^{\gamma_{i+2}},$$

$$\tilde{\delta}'_i = q^{\delta'_i} = q^{\gamma_i}, \qquad \tilde{\delta}'_{i+2} = h(\delta'_{i+2}) = h(\gamma'_{i+2}),$$

and it is still true that $\tilde{\gamma}_i\tilde{\gamma}_{i+2} = \tilde{\delta}'_i\tilde{\delta}'_{i+2}$. This proves (10.3).

Now we prove

$$\prod_{i \text{ odd}} \tilde{\gamma}_i = \prod_{i \text{ odd}} \tilde{\delta}'_i. \tag{10.4}$$

When i is odd, it follows from Lemma 8.2 that γ_i is circled or boxed in the Γ indexing if and only if δ'_i is. Using (9.1), and remembering that since $i - 1$ and $i + 1$ are even we are now assuming γ_{i-1} and γ_{i+1} are multiples of n, we obtain

$$\tilde{\delta}'_i = q^{\gamma_{i-1}-\gamma_{i+1}}\tilde{\gamma}_i.$$

Thus taking the product over odd i, the powers of q will cancel in pairs, giving (10.4). Combining this with (10.3), the theorem is proved.   □

There is another important case where (6.13) is true. This is case where the pattern t is *stable*. We say that t is *stable* if each a_i equals either l_i or l_{i+1}, and each b_i equals either a_i or a_{i+1}. Thus every element of the Γ preaccordion is either circled or boxed. If this is true then it follows from Lemma 8.2 that t' is also stable. Theorem 10.1 does not apply to stable patterns since they usually have critical resonances.

THEOREM 10.2 *Suppose that* t *is stable. Then* $G_\Gamma(t) = G_\Delta(t')$.

Proof. It is easy to see that every element of the Γ and Δ' preaccordions is either circled or boxed, and that the circled entries are precisely the ones that equal zero. As we will explain, the boxed elements are precisely the same for the Γ and Δ' preaccordions.

Let S be the set of elements of the top row of t. Between the top row and the row below it, one element is omitted; call it a. Between this row and the next, another element is omitted; call this b. In t' the same two elements are dropped, but in reverse order. The boxed entries that appear in Γ are

first row:	$\{x - a \mid x \in S, x > a\}$
second row:	$\{b - x \mid x \in S, x < b, x \neq a\}$

The boxed entries that appear in Δ' are:

first row:	$\{b - x \mid x \in S, x > b\}$
second row:	$\{x - a \mid x \in S, x > a, x \neq b\}$

The entry $a - b$ appears in both cases only if $a > b$. The statement is now clear.   □

Chapter Eleven

Types

We now divide the prototypes into much smaller units that we call *types*. We fix a top and bottom row, and therefore a cartoon. For each episode $\mathcal{E}$ of the cartoon, we fix an integer $k_{\mathcal{E}}$. Then the set $\mathfrak{G}$ of all short Gelfand-Tsetlin patterns (6.2) with the given top and bottom rows such that for each $\mathcal{E}$

$$\sum_{\alpha \in \Theta_1 \cap \mathcal{E}} \mathfrak{t}(\alpha) = k_{\mathcal{E}} \tag{11.1}$$

is called a *type*. Thus two patterns are in the same type if and only if they have the same top and bottom rows (and hence the same cartoon), and if the sum of the first (middle) row elements in each episode is the same for both patterns.

Let us choose Γ and Δ' indexings as in Proposition 9.1. With notations as in that Proposition, and $\mathcal{E}$ a fixed episode of the corresponding cartoon, there exist k and l such that $\phi(i) \in \mathcal{E}$ and $\psi(i) \in \mathcal{E}$ precisely when $k \leqslant i \leqslant l$. let

$$L_{\mathcal{E}} = \begin{cases} k & \text{if } k \text{ is even,} \\ k-1 & \text{if } k \text{ is odd,} \end{cases} \qquad R_{\mathcal{E}} = \begin{cases} l & \text{if } l \text{ is even,} \\ l+1 & \text{if } l \text{ is odd.} \end{cases}$$

Then Proposition 9.1 implies that

$$\sum_{i=k}^{l} \delta_i(\mathfrak{t}') = \left(\sum_{i=k}^{l} \gamma_i(\mathfrak{t}) \right) + \gamma_{L_{\mathcal{E}}}(\mathfrak{t}) - \gamma_{R_{\mathcal{E}}}(\mathfrak{t}), \tag{11.2}$$

for all elements of the type. We recall that our convention was that $\gamma_0 = \gamma_{2d+2} = 0$. We take $L_{\mathcal{E}} = 0$ for the first (leftmost) cartoon and $R_{\mathcal{E}} = 2d+2$ for the last episode.

We may classify the possible episodes into four classes generalizing the classification in Table 9.1, and indicate in each case the locations of $\gamma_{L_{\mathcal{E}}}$ and $\gamma_{R_{\mathcal{E}}}$ in the Γ preaccordions, which may be checked by comparison with Table 9.1. Indeed, it must be remembered that in that proof, every panel of type R is replaced by one of type T or type B. Whichever choice is made, Table 9.1 gives the same location for $L_{\mathcal{E}}$ and $R_{\mathcal{E}}$. The classification of the episode into one of four types is given in Table 11.1.

The location of $\delta'_{L_{\mathcal{E}}}$ and $\delta'_{R_{\mathcal{E}}}$ in the Δ preaccordion of $\mathfrak{t}'$ may also be read off from Table 9.1. The classification of the episode into one of four classes is given in Table 11.2.

PROPOSITION 11.1 *If $\mathcal{F}$ is the episode that consecutively follows $\mathcal{E}$, then $R_{\mathcal{E}} = L_{\mathcal{F}}$. The values $\gamma_{L_{\mathcal{E}}}(\mathfrak{t})$ and $\gamma_{R_{\mathcal{E}}}(\mathfrak{t})$ are constant on each type. Moreover $G_{\Gamma}(\mathfrak{t}) = G_{\Delta}(\mathfrak{t}') = 0$ for all patterns $\mathfrak{t}$ in the type unless the $\gamma_{L_{\mathcal{E}}}$ are divisible by n. In the Γ and Δ' preaccordions, $\gamma_{L_{\mathcal{E}}}$ and $\delta'_{L_{\mathcal{E}}}$ may be circled or not, but never boxed.*

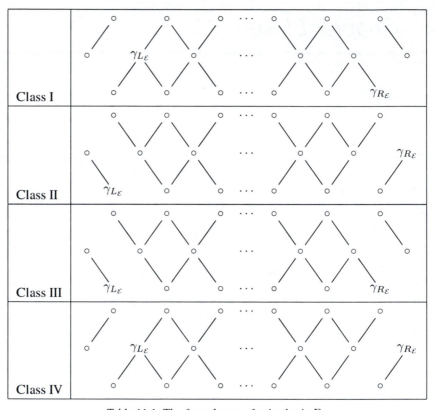

Class I
Class II
Class III
Class IV

Table 11.1 The four classes of episodes in Γ_t.

Proof. From Tables 11.1 and 11.2, it is clear that $R_{\mathcal{E}} = L_{\mathcal{F}}$ for consecutive episodes.

If $\mathcal{E}$ is of Class I or Class IV, then we see that

$$\gamma_{L_{\mathcal{E}}} = \sum_{\mathcal{F} \geqslant \mathcal{E}} \left(\sum_{\alpha \in \Theta_1 \cap \mathcal{F}} t(\alpha) - \sum_{\alpha \in \Theta_0 \cap \mathcal{F}} t(\alpha) \right),$$

where the notation means that we sum over all episodes to the right of $\mathcal{E}$ (including $\mathcal{E}$ itself). If $\mathcal{E}$ is of Class II or III, we have

$$\gamma_{L_{\mathcal{E}}} = \sum_{\mathcal{F} \leqslant \mathcal{E}} \left(\sum_{\alpha \in \Theta_1 \cap \mathcal{F}} t(\alpha) - \sum_{\alpha \in \Theta_2 \cap \mathcal{F}} t(\alpha) \right).$$

In either case, these formulas imply that $\gamma_{L_{\mathcal{E}}}$ is constant on the patterns of the type.

Given their described locations, the fact that $\gamma_{L_{\mathcal{E}}} = \delta'_{L_{\mathcal{E}}}$ is never boxed in either the Γ or Δ' preaccordions may be seen from the definitions.

We now show that $G_{\Gamma}(t) = G_{\Delta}(t') = 0$ unless $n | \gamma_{L_{\mathcal{E}}}$. Indeed, it follows from an examination of the locations of $\gamma_{L_{\mathcal{E}}}$ in the Γ preaccordions and in the Δ' preaccordions (where the same value appears as $\delta'_{L_{\mathcal{E}}}$) that this entry is unboxed in both Γ

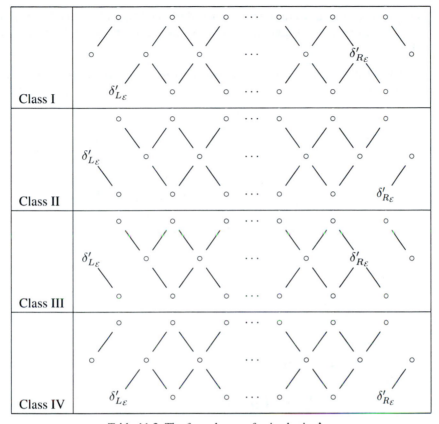

Table 11.2 The four classes of episodes in $\Delta_{t'}$.

and Δ'. If it is uncircled in the Γ preaccordion, then $G_\Gamma(t)$ is divisible by $h(\gamma_{L_\mathcal{E}})$, hence vanishes unless $n|\gamma_{L_\mathcal{E}}$. If it is circled, then we apply the Circling Lemma (Lemma 9.6) to conclude that the same value appears somewhere else uncircled and unboxed, unless $\gamma_{L_\mathcal{E}} = 0$ (which is divisible by n), which again forces $G_\Gamma(t) = 0$ if $n \nmid \gamma_{L_\mathcal{E}}$; and similarly for $G_\Delta(t') = 0$. $\square$

- Due to this result, we may impose the assumption that $n|\gamma_{L\mathcal{E}}$ for every episode. This assumption is in force for the rest of the book.

Now let $\mathcal{E}_1, \cdots, \mathcal{E}_N$ be the episodes of the cartoon arranged from left to right, and let $k_i = k(\mathcal{E}_i)$. By a *local pattern* on $\mathcal{E}_i$ subordinate to $\mathfrak{S}$ we mean an integer-valued function on $\mathcal{E}_i$ that can occur as the restriction of an element of $\mathfrak{S}$ to $\mathcal{E}_i$. Its top and bottom rows are thus the restrictions of the given top rows, and it follows from the definition of the episode that if $(0, t)$ and $(2, t - 1)$ are both in $\mathcal{E}_i$ then $t(0, t) = t(2, t - 1)$; that is, if both an element of the top row and the element of the bottom row that is directly below it are in the same episode, then t has the same value on both, and patterns in the type are resonant at t. The local pattern is subject to the same inequalities as a short pattern, and by (11.2) the sum of its first (middle) row elements must be k_i. Let $\mathfrak{S}_i$ be the set of local patterns subordinate to $\mathfrak{S}$. We call $\mathfrak{S}_i$ a *local type*.

LEMMA 11.2 *A pattern is in $\mathfrak{S}$ if and only if its restriction to $\mathcal{E}_i$ is in $\mathfrak{S}_i$ for each i and so we have a bijection*

$$\mathfrak{S} \cong \mathfrak{S}_1 \times \cdots \times \mathfrak{S}_N.$$

Proof. This is obvious from the definitions, since the inequalities (11.1) for the various episodes are independent of each other. $\square$

Now if t is a short pattern let us define for each episode $\mathcal{E}$

$$G_\Gamma^{\mathcal{E}}(t) = \prod_{\alpha \in \mathcal{E} \cap \Theta_B} \begin{cases} g(\alpha) & \text{if } \alpha \text{ is boxed in } \Gamma_t, \\ q^\alpha & \text{if } \alpha \text{ is circled in } \Gamma_t, \\ h(\alpha) & \text{otherwise,} \end{cases}$$

$$G_\Delta^{\mathcal{E}}(t) = \prod_{\alpha \in \mathcal{E} \cap \Theta_B} \begin{cases} g(\alpha) & \text{if } \alpha \text{ is boxed in } \Delta_t, \\ q^\alpha & \text{if } \alpha \text{ is circled in } \Delta_t, \\ h(\alpha) & \text{otherwise,} \end{cases} \tag{11.3}$$

provided t is *locally strict* at $\mathcal{E}$, by which we mean that if $\alpha, \beta \in \mathcal{E} \cap \Theta_1$ and α is to the left of β then $t(\alpha) > t(\beta)$. If t is not locally strict, then we define $G_\Gamma^{\mathcal{E}}(t) = G_\Delta^{\mathcal{E}}(t) = 0$.

PROPOSITION 11.3 *Suppose that $n|\gamma_{L\mathcal{E}}$ for every episode. Assume also that for each $\mathfrak{S}_i$ we have*

$$\sum_{t_i \in \mathfrak{S}_i} G_\Delta^{\mathcal{E}_i}(t_i') = q^{\gamma_{L\mathcal{E}_i} - \gamma_{R\mathcal{E}_i}} \sum_{t \in \mathfrak{S}_i} G_\Gamma^{\mathcal{E}_i}(t_i). \tag{11.4}$$

Then

$$\sum_{t \in \mathfrak{S}} G_\Delta(t') = \sum_{t \in \mathfrak{S}} G_\Gamma(t). \tag{11.5}$$

This proposition is the bridge between types and local types. Two observations are implicit in the statement of (11.4).

- Since by its definition $G_\Gamma^{\mathcal{E}_i}(t)$ depends only on the restriction t_i of t to $\mathfrak{S}_i$, we may write $G_\Gamma^{\mathcal{E}_i}(t_i)$ instead of $G_\Gamma^{\mathcal{E}_i}(t)$, and this is well-defined.

- The statement uses the fact that $\gamma_{L\mathcal{E}}(t)$ and $\gamma_{R\mathcal{E}}(t)$ are constant on the type, since otherwise $q^{\gamma_{L\mathcal{E}_i} - \gamma_{R\mathcal{E}_i}}$ would be inside the summation.

Proof. If $t_i \in \mathfrak{S}_i$ is the restriction of $t \in \mathfrak{S}$, we have

$$\sum_{t \in \mathfrak{S}} G_\Gamma(t) = \prod_i \sum_{t_i \in \mathfrak{S}_i} G_\Gamma^{\mathcal{E}_i}(t_i) = \prod_i q^{\gamma_{L\mathcal{E}_i} - \gamma_{R\mathcal{E}_i}} \sum_{t_i \in \mathfrak{S}_i} G_\Delta^{\mathcal{E}_i}(t_i').$$

By Proposition 11.1 we have $R_{\mathcal{E}_i} = L_{\mathcal{E}_{i+1}}$. Since our convention is that $\gamma_0 = \gamma_{2d+2} = 0$, it follows that the factors $q^{\gamma_{L\mathcal{E}_i} - \gamma_{R\mathcal{E}_i}}$ cancel. Thus

$$\sum_{t \in \mathfrak{S}} G_\Gamma(t) = \prod_i \sum_{t_i \in \mathfrak{S}_i} G_\Delta^{\mathcal{E}_i}(t_i') = \sum_{t \in \mathfrak{S}} G_\Delta(t').$$

This completes the proof.   □

In the rest of the chapter we will fix an episode $\mathcal{E} = \mathcal{E}_i$, and denote $L = L_\mathcal{E}$ and $R = R_\mathcal{E}$ to simplify the notation. The four remaining Propositions give relations between the Γ and Δ' preaccordions within the episode $\mathcal{E}$.

PROPOSITION 11.4 *Let* t *be a short pattern whose cartoon contains the following Class II resonant episode $\mathcal{E}$ of order d:*

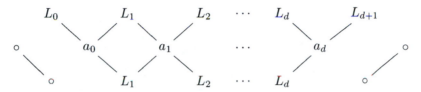

Then there exist integers $s, \mu_1, \nu_1, \mu_2, \nu_2, \cdots, \mu_d, \nu_d$ such that $\mu_i + \nu_i = s$ ($i = 1, \cdots, d$), and the Γ and Δ' preaccordions are given in the following table.

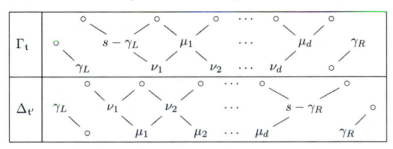

The values s, γ_L and γ_R are constant on the type containing the pattern.

Note: If the episode $\mathcal{E}$ occurs at the left edge of the cartoon, then our convention is that $\gamma_L = \gamma_0 = 0$, and if $\mathcal{E}$ occurs at the right edge of the cartoon, then $\gamma_R = \gamma_{2d+2} = 0$. We would modify the picture by omitting γ_L or γ_R in these cases, but the proof below is unchanged.

Proof. Let $\gamma_L = \gamma_{L_\mathcal{E}}$ and $\gamma_R = \gamma_{R_\mathcal{E}}$ in the notation of the previous chapter, and let s, μ_i, ν_i be defined by their locations in the Γ preaccordion. Let $s = \hat{s} + \gamma_R + \gamma_L$, $\mu_i = \hat{\mu}_i + \gamma_R$ and $\nu_i = \hat{\nu}_i + \gamma_L$. It is immediate from the definitions that

$$\hat{s} = \sum_{j=0}^{d}(a_j - L_{j+1}), \qquad \hat{\mu}_i = \sum_{j=i}^{d}(a_j - L_{j+1}), \qquad \hat{\nu}_i = \sum_{j=0}^{i-1}(a_j - L_{j+1}).$$

From this it we see that $\mu_i + \nu_i = s$ and $\hat{\mu}_i + \hat{\nu}_i = \hat{s}$.

In order to check the correctness of the Δ' diagram, we observe that the resonance contains d panels of type R, each of which may be specialized to a panel of type T or B. We specialize these to panels of type T. We obtain the following canonical snakes, representing the Γ and Δ' preaccordions.

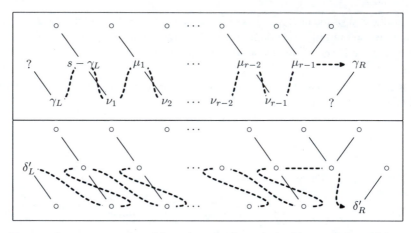

Looking at the even-numbered locations in these indexings, starting with $\gamma_L = \delta'_L$, Proposition 9.1 asserts that the values $\nu_1, \cdots, \nu_d$ are as advertised in the Δ' labeling. It also asserts that the value at the first odd-numbered location, which is the first spot in the bottom row of the episode, is $(s - \gamma_L) + \gamma_L - \nu_1 = \mu_1$; the second odd-numbered location gets the value $\mu_1 + \nu_1 - \nu_2 = \mu_2$, and so forth. $\square$

PROPOSITION 11.5 *Let* t *be a short pattern whose cartoon contains a Class I resonant episode* $\mathcal{E}$ *of order d. There exist integers* $s, \mu_1, \nu_1, \mu_2, \nu_2, \cdots, \mu_d, \nu_d$ *such that* $\mu_i + \nu_i = s$ ($i = 1, \cdots, d$), *and the portions of in* Γ *and* Δ' *preaccordions in* $\mathcal{E}$ *are given in the following table.*

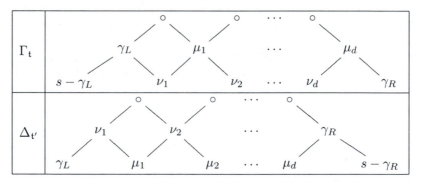

The values $s, \gamma_L,$ *and* γ_R *are constant on the type containing the pattern.*

Proof. We define s, μ_i, and ν_i to be the quantities that make the Γ preaccordion correct. The correctness of the second diagram may be proved using snakes as in Proposition 11.4. The proof that $\mu_i + \nu_i = s$ is also similar to Proposition 11.4. $\square$

PROPOSITION 11.6 *Let* t *be a short pattern whose cartoon contains a Class III resonant episode* $\mathcal{E}$ *of order d. There exist integers* $s, \mu_1, \nu_1, \mu_2, \nu_2, \cdots, \mu_d, \nu_d$ *such that* $\mu_i + \nu_i = s$ ($i = 1, \cdots, d$), *and the portions of the* Γ *and* Δ' *preaccor-*

dions in $\mathcal{E}$ are given in the following table.

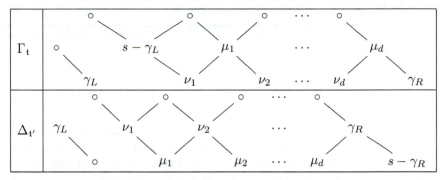

The values s, γ_L, γ_R, and ξ are constant on the type containing the pattern.

Proof. This is similar and left to the reader.   □

PROPOSITION 11.7 *Let t be a short pattern whose cartoon contains a Class IV resonant episode $\mathcal{E}$ of order d. There exist integers $s, \mu_1, \nu_1, \mu_2, \nu_2, \cdots, \mu_d, \nu_d$ such that $\mu_i + \nu_i = s$ $(i = 1, \cdots, d)$, and the portions of the Γ and Δ' preaccordions in $\mathcal{E}$ are given in the following table.*

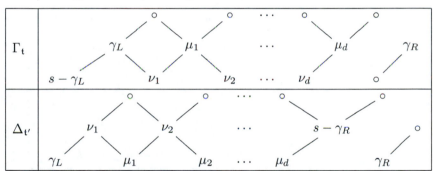

The values s, γ_L, γ_R, and ξ are constant on the type containing the pattern.

Proof. This is similar and left to the reader.   □

Chapter Twelve

Knowability

We turn now to the Knowability Lemma, which explains when products of Gauss sums associated to elements of a preaccordion are explicitly evaluable as polynomials in q, the order of the residue class field. We refer to Chapter 6 for additional discussion of knowability and its role in the proof of Theorem 1.2.

Let $\mathfrak{G} = \prod \mathfrak{G}_i$ be a type. Let $\mathcal{E} = \mathcal{E}_i$ be an episode in the cartoon associated to the short Gelfand-Tsetlin pattern $\mathfrak{t} \in \mathfrak{G}$. If the episode is of Class II, let $a_0, \cdots, a_d$ and $L_1, \cdots, L_{d+1}$ be as in Proposition 11.4. If the class is I, III, or IV, we still define the a_i and L_i analogously:

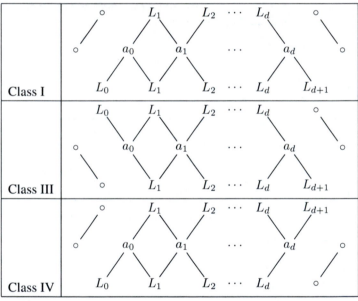

We say that $\mathfrak{t}$ is $\mathcal{E}$-*maximal* if $a_0 = L_0, \cdots, a_d = L_d$, and $\mathcal{E}$-*minimal* if $a_0 = L_1, \cdots, a_d = L_{d+1}$. Not every local type $\mathfrak{G}_i$ (with $\mathcal{E} = \mathcal{E}_i$) contains an $\mathcal{E}$-maximal or $\mathcal{E}$-minimal element. If it does, then $\mathfrak{G}_i$ consists of that single local pattern.

PROPOSITION 12.1 *If* $\mathfrak{t}_i$ *is* $\mathcal{E}_i$-*maximal then* $\mathfrak{t}'_i$ *is* $\mathcal{E}_i$-*minimal, and*

$$G_\Delta^{\mathcal{E}_i}(\mathfrak{t}'_i) = q^{\gamma_{L_{\mathcal{E}_i}} - \gamma_{R_{\mathcal{E}_i}}} G_\Gamma^{\mathcal{E}_i}(\mathfrak{t}_i).$$

Proof. Let notations be as in Proposition 11.4, 11.5, 11.6, or 11.7, depending on the class of $\mathcal{E} = \mathcal{E}_i$, and in particular $L = L_{\mathcal{E}_i}$ and $R = R_{\mathcal{E}_i}$. Each entry in the $\mathcal{E}$-portion of both $\Gamma_\mathfrak{t}$ and $\Delta_{\mathfrak{t}'}$ is boxed except γ_L and γ_R, if they happen to lie inside $\mathcal{E}$, which one or both does unless $\mathcal{E}$ is of Class II; these are neither boxed nor circled. Therefore $G_\Gamma^{\mathcal{E}_i}(t_i)$ equals

$$g(s-\gamma_L)\prod g(\mu_i)g(\nu_i)\times\left\{\begin{array}{ll} h(\gamma_L) & \text{Class I or IV} \\ 1 & \text{Class II or III} \end{array}\right. \times\left\{\begin{array}{ll} h(\gamma_R) & \text{Class I or III} \\ 1 & \text{Class II or IV} \end{array}\right.$$

and $G_\Delta^{\mathcal{E}_i}(t')$ is the same, except that $g(s - \gamma_L)$ is replaced by $g(s - \gamma_R)$. By Proposition 11.1 we may assume as usual that $n|\gamma_L$ and $n|\gamma_R$. By Proposition 8.1 it follows that $q^{\gamma_L - \gamma_R}g(s - \gamma_L) = g(s - \gamma_R)$, and the statement is proved. $\qquad\square$

PROPOSITION 12.2 (Knowability Lemma) *Let $\mathcal{E}$ be an episode in the cartoon associated to the short Gelfand-Tsetlin pattern* t, *and let $L = L_\mathcal{E}$ and $R = R_\mathcal{E}$ as in Tables 11.1 and 11.2. Let s, μ_i and ν_i be as in Proposition 11.4, 11.5, 11.6, or 11.7, depending on the class of $\mathcal{E}$. Assume that $n \nmid s$. Then either $G_\Gamma(t) = G_\Delta(t') = 0$, or* t *is $\mathcal{E}$-maximal.*

The term "Knowability Lemma" should be understood as follows. It asserts that one of the following cases applies:

- Maximality: t is $\mathcal{E}$-maximal, and $\mathfrak{S}_i$ consists of the single local pattern. In this case (11.4) follows from Proposition 12.1.

- Knowability: $n|s$ in which case all the Gauss sums that appear in all the patterns of the resotope appear in knowable combinations — $g(s)$ by itself or $g(\mu_i)g(\nu_i)$ where $\mu_i + \nu_i = s$.

- In all other cases where $n \nmid s$ we have and $G_\Gamma(t) = G_\Delta(t') = 0$ for all patterns so (11.4) is obvious.

Knowability (as explained in Chapter 6) *per se* is not important for the proof that Statement C implies Statement B, but the precise statement in Proposition 12.2, particularly the fact that we may assume that $n|s$, *will* be important. Theorems 15.3 and 15.4 below validate the term "knowability" by explicitly evaluating the sums that arise when $n|s$.

Proof. We will discuss the cases where $\mathcal{E}$ is Class II or Class I, leaving the remaining two cases to the reader.

First assume that $\mathcal{E}$ is Class II. Let notations be as in Proposition 11.4. By Proposition 11.1 we may assume that $n|\gamma_L$ and $n|\gamma_R$. We will assume that $G_\Gamma(t) \neq 0$ and show that $s - \gamma_L, \mu_i$, and ν_i are all boxed. A similar argument would give the same conclusion assuming $G_\Delta(t') \neq 0$. Since $h(x) = 0$ when $n \nmid x$, if such x appears in Γ_t it is either boxed or circled. In particular $s - \gamma_L$ is either boxed or circled.

We will argue that $s - \gamma_L$ is not circled. By the Circling Lemma (Lemma 9.6), $\mu_1 = s - \gamma_L, \nu_1 = \gamma_L$, and ν_1 is also circled. Now $n \nmid \mu_1 = s - \gamma_L$, so μ_1 is either circled or boxed, and it cannot be boxed, because this would imply that ν_1 is both circled and boxed, which is impossible since $G_\Gamma(t) \neq 0$. Thus $\mu_1 = s - \gamma_L$ is circled, and we may repeat the argument, showing that $s - \gamma_L = \mu_1 = \mu_2 = \dots$ so that $\nu_1 = \nu_2 = \dots$, and that all entries are circled. When we reach the end of the top row, μ_d is circled, which implies that $\mu_d = 0$, and so $s = \gamma_L$, which is a contradiction since we assumed that $n \nmid s$.

This proves that $s - \gamma_L$ is boxed. Now we argue by contradiction that the μ_i and ν_i are also boxed. If not, let $i \geqslant 0$ be chosen so that $\nu_1, \cdots, \nu_{i-1}$ are boxed (and therefore, so are $\mu_1, \cdots, \mu_{i-1}$) but ν_i is not boxed. We note that ν_i cannot be circled, because if ν_i is circled, then μ_{i-1} (or s if $i = 0$) is both circled and boxed, which is a contradiction. Thus ν_i is neither boxed nor circled and so $n | \nu_i$. Since $\nu_i + \mu_i = s$ and $n \nmid s$, we have $n \nmid \mu_i$ and so μ_i is either boxed or circled. It cannot be boxed since this would imply that ν_i is also boxed, and our assumption is that it is not. Thus μ_i is circled. By the Circling Lemma, $\mu_i = \mu_{i+1}$, and so $n \nmid \mu_{i+1}$, which is thus either boxed or circled. It cannot be circled, since if it is, then ν_{i+1} is both circled (since μ_i is circled) and boxed (since μ_{i+1} is boxed), and we know that if a bottom row entry is both boxed and circled, then t is not strict and $G_\Gamma(t) = 0$, which is a contradiction. Thus μ_{i+1} is circled. Repeating this argument, $\mu_i = \mu_{i+1} = \cdots$ are all circled, and when we get to the end, μ_d is circled, so by the Circling Lemma, $\mu_i = \mu_d = \gamma_R$, which is a contradiction since γ_R is divisible by n, but μ_i is not. This contradiction shows that $s - \gamma_L$ and the μ_i, ν_i are all boxed, and it follows from the definitions that t is $\mathcal{E}$-maximal.

We now discuss the variant of this argument for the case that $\mathcal{E}$ is of Class I, leaving the two other cases to the reader. Let notations be as in Proposition 11.5. Again we assume that $G_\Gamma(t) \neq 0$, so whenever x appears in Γ_t with $n \nmid x$ it is either boxed or circled. Due to its location in the cartoon, there is no way that $s - \gamma_L$ can be circled, so it is boxed.

Now we argue by contradiction that $\nu_1, \cdots, \nu_d$ and hence $\mu_1, \cdots, \mu_d$ are all boxed. If not, let $i \geqslant 0$ be chosen so that $\nu_1, \cdots, \nu_{i-1}$ are boxed (and therefore, so are $\mu_1, \cdots, \mu_{i-1}$) but ν_i is not boxed. The same argument as in the Class II case shows that $n | \nu_i$ so $n \nmid \mu_i$ and that μ_i is circled, and moreover that $\mu_i = \mu_{i+1} = \cdots = \mu_d$ and that these are all circled. But now this is a contradiction since due to its location in the cartoon, μ_d cannot be circled. $\qquad\square$

Chapter Thirteen

The Reduction to Statement D

We now switch to the language of resotopes, as defined in Chapter 6. We remind the reader that we may assume $\gamma_{L_\mathcal{E}}$ and $\gamma_{R_\mathcal{E}}$ are multiples of n for every totally resonant episode. We also recall that s and d are the weights of the accordions (6.15) and (6.17) under consideration.

PROPOSITION 13.1 *Statement D is equivalent to Statement C. Moreover, Statement D is true if $n \nmid s$.*

Proof. The case of a totally resonant short Gelfand-Tsetlin pattern $\mathfrak{t}$ is a special case of Proposition 11.4, and the point is that $\Gamma_\mathfrak{t}$ is a Γ-accordion $\mathfrak{a}$, and Proposition 11.4 shows that $\Delta_{\mathfrak{t}'}$ is the Δ-accordion $\mathfrak{a}'$. In this case $\gamma_L = \gamma_R = 0$. Moreover as $\mathfrak{t}$ runs through its totally resonant prototype, $\mathfrak{a}$ runs through the Γ-resotope $\mathcal{A}_s(c_0, \cdots, c_d)$ with $c_i = L_i - L_{i+1}$, so Statement D boils down to Statement C. The fact that Statement D is true when $n \nmid s$ follows from the Knowability Lemma and Proposition 12.1. $\quad\square$

We turn next to the proof that Statement D implies Statement B. What we will show is that for each of the four types of resonant episodes, Statement D implies (11.4); then Statement B will follow from Proposition 11.3. We fix an episode $\mathcal{E} = \mathcal{E}_i$, and will denote $L = L_i$, $R = R_i$. By Proposition 11.1 we may assume that $n|\gamma_L$ and $n|\gamma_R$. Moreover by the Knowability Lemma (Proposition 12.2) we may assume $n|s$, where s and other notations are as in Proposition 11.4, 11.5, 11.6 or 11.7, depending on the class of $\mathcal{E}$.

PROPOSITION 13.2 *Let $\mathcal{E}$ be a Class II episode, and let notations be as in Proposition 11.4. If*

$$
\mathfrak{a} = \left\{ \begin{matrix} \hat{s} & & \hat{\mu}_1 & & \cdots & & \hat{\mu}_d \\ & \hat{\nu}_1 & & & \cdots & & \hat{\nu}_d \end{matrix} \right\}
$$

where $s = \hat{s} + \gamma_R + \gamma_L$, $\mu_i = \hat{\mu}_i + \gamma_R$ and $\nu_i = \hat{\nu}_i + \gamma_L$, then $\mathfrak{a}$ lies in the resotope $\mathcal{A} = \mathcal{A}_{\hat{s}}(c_0, \cdots, c_d)$ with $c_i = L_i - L_{i+1}$; let σ denote the signature of $\mathfrak{a}$ in $\mathcal{A}$. Then $\mathfrak{t} \longmapsto \mathfrak{a}_\sigma$ induces a bijection from the local type $\mathfrak{S}_i$ to $\mathcal{A}$. Assume furthermore that $n|\gamma_L$ and $n|\gamma_R$. Then

$$
q^{\gamma_L} G_\Gamma^\mathcal{E}(\mathfrak{t}) = q^{(d+1)(\gamma_R + \gamma_L)} \mathcal{G}_\Gamma(\mathfrak{a}_\sigma), \qquad q^{\gamma_R} G_\Delta^\mathcal{E}(\mathfrak{t}') = q^{(d+1)(\gamma_R + \gamma_L)} \mathcal{G}_\Delta(\mathfrak{a}_\sigma').
$$
(13.1)

Proof. With notations as in Proposition 11.4, the inequalities $L_i \geqslant a_i \geqslant L_{i+1}$ that a_i must satisfy can be written (with $\hat{\mu}_0 = \hat{s}$):

$$
L_i - L_{i+1} \geqslant \hat{\mu}_{i-1} - \hat{\mu}_i \geqslant 0,
$$

which are the same as the conditions that $\mathfrak{a}_\sigma$ lies in $\mathcal{A} = \mathcal{A}_s(c_0, \cdots, c_d)$, with $c_i = L_i - L_{i+1}$. Each entry in $\mathfrak{a}_\sigma$ is boxed or circled if and only if the corresponding entry in Γ_t is, and similarly, every entry in $\mathfrak{a}'_\sigma$ is boxed or circled if and only if the corresponding entry in the (left-to-right) mirror image of $\Delta_{t'}$ is. Using the assumption that $n|\gamma_R$ and $n|\gamma_L$ and Proposition 8.1 we can pull a factor of q^{γ_R} from the factor of $G_\Gamma^\mathcal{E}(t)$ corresponding to $s - \gamma_L = \hat{s} + \gamma_R$, which is

$$\begin{cases} g(\hat{s} + \gamma_R) & \text{if } s - \gamma_L \text{ is boxed;} \\ q^{\hat{s} + \gamma_R} & \text{if } s - \gamma_L \text{ is circled;} \\ h(\hat{s} + \gamma_R) & \text{otherwise,} \end{cases}$$

leaving just the corresponding contribution in $\mathcal{G}_\Gamma(\mathfrak{a}_\sigma)$; and similarly we may pull out d factors of q^{γ_R} from the contributions of $\mu_i = \hat{\mu}_i + \gamma_R$, and d factors of q^{γ_L} from the contributions of $\nu_i = \hat{\nu}_i + \gamma_L$. What remains is just $\mathcal{G}_\Gamma(\mathfrak{a}_\sigma)$. This gives the first identity in (13.1), and the second one is proved similarly. $\qquad \square$

Corollary to Proposition 13.2. *Statement D implies (11.4) for Class II episodes.*

Although this reduction was straightforward for Class I, each of the remaining classes involves some nuances. In every case we will argue by comparing $q^{\gamma_L} G_\Gamma(t)$ to $\mathcal{G}_\Gamma(\mathfrak{a}_\sigma)$, where $\mathfrak{a}_\sigma$ is the accordion associated with the totally resonant pattern

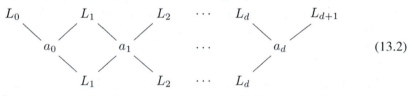

$$(13.2)$$

Here L_i and a_i are as in Proposition 11.4, 11.5, 11.6, or 11.7. We have moved L_0 and L_d from the bottom row to the top row as needed, and discarded the rest of the top and bottom rows. The notations s, μ_i, and ν_i are already in use in Proposition 11.4, 11.5, 11.6, or 11.7, so we will denote

$$\mathfrak{a} = \left\{ \begin{array}{ccccc} t & & \psi_1 & \cdots & \psi_d \\ & \phi_1 & & \cdots & \phi_d \end{array} \right\}. \qquad (13.3)$$

We see that $\mathfrak{a}_\sigma$ runs through $\mathcal{A}_t(c_0, \cdots, c_d)$ by Proposition 13.2, with $c_i = L_i - L_{i+1}$. We will compare $G_\Gamma(t)$ and $G_\Delta(t')$ with $\mathcal{G}_\Gamma(\mathfrak{a}_\sigma)$ and $\mathcal{G}_\Delta(\mathfrak{a}'_\sigma)$, respectively. A complication is that while corresponding entries of Γ_t and $\mathfrak{a}_\sigma$ are boxed together, the circlings may not quite match; the argument will justify moving circles from one entry in $G_\Gamma(t)$ to another. Specifically, if either γ_L or γ_R is within $\mathcal{E}$ and is circled, the circle needs to be moved to another location. This is justified by the following observation.

LEMMA 13.3 (Moving Lemma) *Suppose that x and y both appear in the $\mathcal{E}$ part of Γ_t, and that y is circled, but x is neither circled nor boxed. Suppose that both x and y are both positive and $x \equiv y$ modulo n. Then we may move the circle from y to x without changing the value of $G_\Gamma^\mathcal{E}(t)$.*

Proof. Before moving the circle, the contribution of the two entries is $q^y h(x)$; after moving the circle, the contribution is $q^x h(y)$. These are equal by Proposition 8.1. (The positivity of x is needed since $h(0)$ is undefined.) $\qquad \square$

In each case we will discuss $G_\Gamma(t)$ carefully leaving $G_\Delta(t')$ more or less to the reader. The case where $\mathcal{E}$ is of Class II has already been handled in Proposition 13.2.

Class I Episodes

We assume that the $\mathcal{E}$-portion of t has the form:

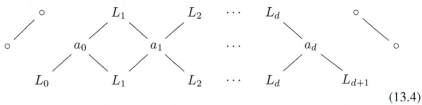

$$(13.4)$$

We will compare $G_\Gamma(t)$ with $\mathcal{G}_\Gamma(\mathfrak{a}_\sigma)$, where $\mathfrak{a}$ is the accordion (13.3) derived from the pattern in (13.2), and σ is its signature. Thus we move L_0 and L_{d+1} to the top row, which does not affect the inequalities that the a_i satisfy, and discard the rest of the pattern to obtain the totally resonant pattern (13.2), then compute its accordion. Otherwise, let Γ_t and $\Delta_{t'}$ be as in Proposition 11.5, and let s, μ_i, and ν_i be as defined there.

PROPOSITION 13.4 *Assume $\mathcal{E}$ is a Class I episode and that $n|s, \gamma_L, \gamma_R$. As t runs through its local type, $\mathfrak{a}_\sigma$ runs through $\mathcal{A}_t(c_0, \cdots, c_d)$ with $c_i = L_i - L_{i+1}$, and*

$$q^{\gamma_L} G^{\mathcal{E}}_\Gamma(t) = h(\gamma_L)h(\gamma_R)q^{(d+1)(2s-\gamma_L-\gamma_R)}\mathcal{G}_\Gamma(\mathfrak{a}_\sigma),$$
$$q^{\gamma_R} G^{\mathcal{E}}_\Delta(t') = h(\gamma_L)h(\gamma_R)q^{(d+1)(2s-\gamma_L-\gamma_R)}\mathcal{G}_\Delta(\mathfrak{a}'_\sigma).$$

Proof. Using (13.4) and (13.2), we have

$$t = \sum_{j=0}^{d}(a_j - L_{j+1}) = \gamma_R - (s - \gamma_L), \qquad \psi_i = \sum_{j=i}^{d}(a_j - L_{j+1}) = \gamma_R - \nu_i,$$

and $\phi_i + \psi_i = t$. If γ_L is circled, we will move the circle to $s - \gamma_L$. To justify the use of the Moving Lemma (Lemma 13.3) we check that γ_L and $s - \gamma_L$ are both positive and congruent to zero modulo n. Positivity of γ_L follows since $\gamma_L \geqslant \mu_d$, and $\mu_d > 0$ since if $\mu_d = 0$ then it is circled, which it cannot be due to its location in the cartoon. To see that $s - \gamma_L > 0$, if it is zero then both $s - \gamma_L$ and γ_L are circled, which implies that $L_1 = a_1 = L_2$, but $L_1 > L_2$. Both $s - \gamma_L$ and γ_L are multiples of n by assumption.

If γ_R is circled, we will move the circle to μ_d. To see that this is justified, we must check that γ_R and μ_d are positive and multiples of n. We are assuming $n|\gamma_R$, and it is positive since $\gamma_R \geqslant s - \gamma_L$, which cannot be zero; if it were, it would be circled, which it cannot be due to its position in the cartoon. Also by the Circling Lemma, since γ_R is circled it equals ν_d; thus $\mu_d = s - \nu_d = s - \gamma_R \equiv 0$ modulo n. And μ_d cannot be zero since it is not circled, due to its location in the cartoon.

With these circling modifications, γ_L and γ_R are neither circled nor boxed, hence produce factors in $G^{\mathcal{E}}_\Gamma(t)$ of $h(\gamma_L)$ and $h(\gamma_R)$. The remaining factors in $G^{\mathcal{E}}_\Gamma(t)$ can be handled as follows. Let $F(x) = q^x$ if x is a boxed entry in Γ_t or $\mathfrak{a}_\sigma$, $g(x)$ if it is circled, and $h(x)$ if it is neither boxed nor circled. We have

$$q^{\gamma_L} F(s - \gamma_L) = q^{2s - \gamma_L - \gamma_R} F(t),$$
$$F(\mu_i) = q^{s - \gamma_R} F(\psi_i),$$
$$F(\nu_i) = q^{s - \gamma_L} F(\phi_i),$$

and multiplying these identities together gives the stated identity for $q^{\gamma_L} G_\Gamma^{\mathcal{E}}(t)$. (There are two entries $h(\gamma_L)$ and $h(\gamma_R)$ that have to be taken out.) The Δ' preaccordion is handled similarly. $\qquad\square$

Class III Episodes

Now we assume that the $\mathcal{E}$-portion of t has the form:

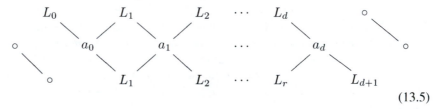

$$(13.5)$$

PROPOSITION 13.5 *Assume $\mathcal{E}$ is a Class III episode and that $n|s, \gamma_L, \gamma_R$. If $a_0 = L_1$, $a_1 = L_2, \cdots, a_d = L_{d+1}$ then the local type consists of a single pattern t, for which (11.4) is satisfied. Assume that this is not the case. Then as t runs through its local type, $\mathfrak{a}_\sigma$ runs through $\mathcal{A}_t(c_0, \cdots, c_d)$ with $c_i = L_i - L_{i+1}$, and*

$$q^{\gamma_L} G_\Gamma^{\mathcal{E}}(t) = q^{(d+1)(s+\gamma_L-\gamma_R)} h(\gamma_R) \mathcal{G}_\Gamma(\mathfrak{a}_\sigma),$$
$$q^{\gamma_R} G_\Delta^{\mathcal{E}}(t') = q^{(d+1)(s+\gamma_L-\gamma_R)} h(\gamma_R) \mathcal{G}_\Delta(\mathfrak{a}'_\sigma).$$

Proof. If $a_0 = L_1$, $a_1 = L_2, \cdots, a_d = L_{d+1}$ then the local type consists of a single element t. We will handle this case separately. For this t it is easy to see that all entries except μ_d are circled in Γ_t, while in $\Delta_{t'}$ all entries except $s - \gamma_R$ are circled. But by the Circling Lemma $s - \gamma_L = \mu_1 = \cdots = \mu_d$ and $\mu_d > 0$ since it cannot be circled due to its location in the cartoon. Thus we may move the circle from $s - \gamma_L$ to μ_d and then compare $G_\Gamma^{\mathcal{E}}(t)$ and $G_\Delta^{\mathcal{E}}(t')$ to see directly that (11.4) is true.

We exclude this case and assume that at least one of the inequalities $a_i \geqslant L_{i+1}$ is strict. Using (13.3) and (13.5) we have $t = \gamma_R - \gamma_L$, $\phi_i = \gamma_R - \gamma_L$, $\psi_i = \nu_i - \gamma_L$, and $\psi_i = \mu_i + \gamma_R - s$, where $\gamma_R, \gamma_L, s, \mu_i$, and ν_i are as in Proposition 11.6, and σ is the signature of $\mathfrak{a}$ in $\mathcal{A}$. If γ_L is circled then we move the circle from γ_L to μ_d in Γ_t. This is justified as in the Class I case, except that the justification we gave there for the claim that $\gamma_R > 0$ is no longer valid. It follows now from our assumption that one of the inequalities $a_i \geqslant L_{i+1}$ is strict. After moving the circle from γ_L to μ_d in Γ_t, each factor $s - \gamma_L$, μ_i, ν_i is circled or boxed in the (circling-modified) Γ_t if and only if the corresponding factor t, ψ_i, or ϕ_i is circled or boxed in $\mathfrak{a}_\sigma$. Moreover $s + \gamma_L - \gamma_R \equiv 0$ modulo n so we can pull out a factor of $q^{s+\gamma_L-\gamma_R}$ from the contributions of $s - \gamma_L$ and each pair μ_i, ν_i, to $q^{\gamma_L} G_\Gamma^{\mathcal{E}}(t)$, and what remains is $h(\gamma_R) \mathcal{G}_\Gamma(\mathfrak{a}_\sigma)$. A similar treatment gives the other identity. $\qquad\square$

Class IV Episodes

Now we assume that the $\mathcal{E}$-portion of t has the form:

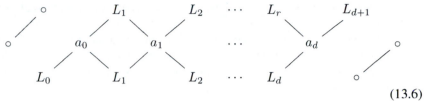

$$(13.6)$$

If $a_0 = L_1, a_1 = L_2, \cdots, a_d = L_{d+1}$ then the local type consists of a single element t. In this case γ_L is circled in both Γ_t and $\Delta_{t'}$ and we don't try to move it. We have

$$q^{\gamma_L} G_\Gamma^{\mathcal{E}}(\mathfrak{t}) = h(s)q^{\gamma_L} \prod_{i=1}^{d} q^{\mu_i} \prod_{i=1}^{d} q^{\nu_i} = h(s)q^{\gamma_L} q^{ds} = q^{\gamma_R} G_\Delta^{\mathcal{E}}(\mathfrak{t}').$$

We exclude this case and assume that at least one of the inequalities $a_i \geqslant L_{i+1}$ is strict. Using (13.3) and (13.6) we have $t = \gamma_L - \gamma_R$, $\psi_i = \mu_i - \gamma_R$, and $\phi_i = \nu_i + \gamma_L - s$.

PROPOSITION 13.6 *Assume $\mathcal{E}$ is a Class IV episode and that $n|s, \gamma_L, \gamma_R$. As t runs through its local type, $\mathfrak{a}_\sigma$ runs through $\mathcal{A}_t(c_0, \cdots, c_d)$ with $c_i = L_i - L_{i+1}$, and*

$$q^{\gamma_L} G_\Gamma^{\mathcal{E}}(\mathfrak{t}) = h(\gamma_L)q^{(d+1)(\gamma_R - \gamma_L + s)} \mathcal{G}_\Gamma(\mathfrak{a}_\sigma),$$
$$q^{\gamma_R} G_\Delta^{\mathcal{E}}(\mathfrak{t}') = h(\gamma_L)q^{(d+1)(\gamma_R - \gamma_L + s)} \mathcal{G}_\Delta(\mathfrak{a}'_\sigma).$$

Proof. If γ_L is circled, we must move the circle from γ_L to $s - \gamma_L$. This is justified the same way as in the Class I case, except that the positivity of γ_L must be justified differently. In this case, it follows from our assumption that one of the inequalities $a_i \geqslant L_{i+1}$ is strict. Now we can pull out a factor of $q^{s + \gamma_R - \gamma_L}$ from the contributions of $s - \gamma_L$ and each pair μ_i, ν_i, and the statement follows as in our previous cases. $\square$

THEOREM 13.7 *Statement D (or, equivalently, Statement C) implies Statement B.*

Proof. The equivalence of Statements D and C is the Corollary to Proposition 13.2. By Proposition 11.3 we must show (11.4) for every episode $\mathcal{E}$. By Proposition 11.1 we may assume that $n|\gamma_L$ and $n|\gamma_R$. Moreover by the Knowability Lemma (Proposition 12.2) we may assume $n|s$ because if $n \nmid s$ then Proposition 12.1 applies. We may then apply Proposition 13.2, 13.4, 13.5, or 13.6 depending on the class of $\mathcal{E}$. $\square$

In the remainder of the chapter, we verify Statement D in a specific example and highlight some of the difficulties in obtaining a general proof. This material is not strictly necessary for the sequel, as the methods used are not directly applied in any of the later reductions or proofs. However, the ideas do motivate some of our later efforts.

A distinguished role in the computation is played by a set of identities involving the functions g and h as defined in (1.15). Let s be a fixed positive integer. We will consider pairs (α_1, α_2), (β_1, β_2), etc. of integers such that $\alpha_1 + \alpha_2 = \beta_1 + \beta_2 = s$. Then we will write $h_\alpha = h(\alpha_1)h(\alpha_2)$, $g_\alpha = g(\alpha_1)g(\alpha_2)$, and so forth.

PROPOSITION 13.8 *Suppose that* $s = \alpha_1 + \alpha_2$. *Then*

$$g(s)h_\alpha - h(s)g_\alpha = h(s)^2 - q^s h(s) = g(s)h(s) \tag{13.7}$$

and

$$h_\alpha h(s) = h_\alpha(g(s) + q^s). \tag{13.8}$$

Proof. We will use Proposition 8.1 without comment in this proof. We have $h(s) = h_\alpha = 0$ unless $n|s$. Therefore we may assume $n|s$. By Proposition 8.1

$$g(s) = -q^{s-1}, \qquad h(s) = (q-1)q^{s-1}.$$

Now there are two cases. If $n|\alpha_1, \alpha_2$ then

$$h_\alpha = (q-1)^2 q^{s-2}, \qquad g_\alpha = q^{s-2}.$$

On the other hand if $n \nmid \alpha_1$ then $n \nmid \alpha_2$ also and

$$h_\alpha = 0, \qquad g_\alpha = q^{s-1}.$$

In either case, the identities may now be checked. $\square$

We now proceed to the example, verifying Statement D in this special case. The example has been chosen as typical of the sorts of interesting behavior that arise in the resonant case. It also appeared in [15], but in different notation, for here we use the notation of resohedra. Let us denote

$$\Upsilon(\mathfrak{p}) = \mathcal{G}_\Gamma(\mathfrak{p}) - \mathcal{G}_\Delta(\mathfrak{p}').$$

Let us call a subset Π of the resotope $\mathcal{A}$ a *packet* if $\sum_{\mathfrak{p} \in \Pi} \Upsilon(\mathfrak{p}) = 0$. For example, if $\mathfrak{p}$ is in the interior of the resotope, then $\Upsilon(\mathfrak{p}) = 0$ by Theorem 10.1 and so $\{\mathfrak{p}\}$ is a singleton packet.

Our approach in the following example is to try to organize the elements of a resotope $\mathcal{A}$ into a disjoint union of packets. If this can be done then Statement D follows. An examination of examples shows that this can be done with fairly small packets. Ultimately this approach is too naive, for the packets are not canonical and there seems to be no simple way of describing a choice of packets.

Therefore we will use this approach in this example, but then introduce new ideas in order to prove Theorem 1.2. Instead of directly constructing packets as we do in this example, we will reduce Statement D to Statement E based on replacing the sum of the $\mathcal{G}_\Gamma$ over the (geometrically complicated) resohedron with a sum of the alternating sums Λ_Γ over the (geometrically simple) cut-and-paste simplex

$\mathrm{CP}_\eta(c_0, \cdots, c_d)$. In the latter context, the packet phenomenon reappears but the difficulties are tamed by the change of viewpoint.

Turning to the example, we consider $\mathcal{A} = \mathcal{A}_s(c_0, c_1, c_2)$ where we assume that

$$c_0 + c_1, c_0 + c_2, c_1 + c_2 < s < c_0 + c_1 + c_2. \tag{13.9}$$

We recall that $\mathcal{A}_s(c_0, c_1, c_2)$ is the set of accordions with top row $\{s, \mu_1, \mu_2\}$ such that

$$0 \leqslant s - \mu_1 \leqslant c_0, \qquad 0 \leqslant \mu_1 - \mu_2 \leqslant c_1, \qquad 0 \leqslant \mu_2 \leqslant c_2.$$

Thus $\mathcal{A}$ is bounded by some of the six lines corresponding to the equalities

$$s - \mu_1 = 0, \qquad s - \mu_1 = c_0, \qquad \mu_1 - \mu_2 = 0,$$

$$\mu_1 - \mu_2 = c_1, \qquad \mu_2 = 0, \qquad \mu_2 = c_2. \tag{13.10}$$

Only those equalities that can actually be attained with the given choices of c_0, c_1, c_2 describe lines bounding the resotope. These are found in the next Proposition.

PROPOSITION 13.9 *The assumptions (13.9) imply that the equalities $\mu_1 - \mu_2 = 0$ and $\mu_2 = 0$ are never attained, but that the remaining four inequalities in (13.10) all are attained. Thus two of the inqualities describing $\mathcal{A}_s(c_0, c_1, c_2)$ are superfluous, and $\mathcal{A}_s(c_0, c_1, c_2)$ is the region bounded by the four lines*

$$s - \mu_1 = 0, \qquad s - \mu_1 = c_0, \qquad \mu_1 - \mu_2 = c_1, \qquad \mu_2 = c_2.$$

Proof. We have

$$\mu_1 - \mu_2 = s - (s - \mu_1) - \mu_2 \geqslant s - c_0 - c_2$$

and

$$\mu_2 = s - (s - \mu_1) - (\mu_1 - \mu_2) \geqslant s - c_0 - c_1,$$

so $\mu_1 - \mu_2$ and μ_2 are both bounded below by positive constants. The inequalities $\mu_1 - \mu_2 = c_1$ and $s - \mu_1 = c_0$ are both attained for the accordion

$$\mathfrak{a} = \left\{ \begin{matrix} s & & s - c_0 & & s - c_0 - c_1 \\ & c_0 & & c_0 + c_1 & \end{matrix} \right\},$$

and the inequalities $\mu_2 = c_2$ and $s - \mu_1 = 0$ are both attained for the accordion

$$\mathfrak{c} = \left\{ \begin{matrix} s & & s & & c_2 \\ & 0 & & s - c_2 & \end{matrix} \right\}.$$

□

Thus $\mathcal{A}_s(c_0, c_1, c_2)$ is represented by the set of lattice points in the (closed) trapezoid:

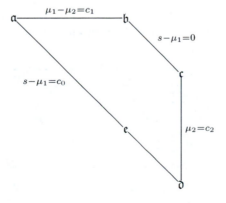

We have already noted that interior points form singleton packets. On the boundary, we will find one packet of cardinality five, and other packets of cardinalities 2 and 3.

We have marked five points on the boundary. To describe these, we introduce pairs $\alpha = (\alpha_1, \alpha_2)$, $\beta = (\beta_1, \beta_2)$, $\gamma = (\gamma_1, \gamma_2)$, $\delta = (\delta_1, \delta_2)$. These parameters are determined by the equations $\alpha_1 = c_0$, $\beta_1 = c_0 + c_1$, $\gamma_1 = c_1$, $\delta_1 = c_2$ together with $\alpha_1 + \alpha_2 = \beta_1 + \beta_2 = \gamma_1 + \gamma_2 = \delta_1 + \delta_2 = s$. Here are the five accordions labeled in the preceding figure.

We may use Proposition 13.8 to simplify the expression for $\Upsilon(\mathfrak{p})$ at each of the five marked points, arriving at the following values.

$\mathfrak{p}$	$\mathcal{G}_\Gamma(\mathfrak{p})$	$\mathcal{G}_\Delta(\mathfrak{p}')$	$\Upsilon(\mathfrak{p})$
$\mathfrak{a}$	$g(s)g_\alpha h_\beta$	$h(s)g_\alpha g_\beta$	$g(s)h(s)g_\alpha$
$\mathfrak{b}$	0	$h(s)^2 g_\gamma$	$-h(s)^2 g_\gamma$
$\mathfrak{c}$	$q^s h(s)g_\delta$	$g(s)h(s)h_\delta$	$-h(s)^2 g(s) - g(s)h(s)g_\delta$
$\mathfrak{d}$	$g(s)h_\alpha g_\delta$	$g(s)g_\alpha h_\delta$	$g(s)h(s)g_\delta - g(s)h(s)g_\alpha$
$\mathfrak{e}$	$g(s)h_\alpha h_\gamma$	$h(s)g_\alpha h_\gamma$	$g(s)h(s)h_\gamma$

Now we have

$$\Upsilon(\mathfrak{a}) + \Upsilon(\mathfrak{d}) + \Upsilon(\mathfrak{c}) = -h(s)^2 g(s),$$
$$\Upsilon(\mathfrak{b}) + \Upsilon(\mathfrak{e}) = h(s)[g(s)h_\gamma - h(s)g_\gamma] = h(s)^2 g(s).$$

Thus $\{\mathfrak{a}, \mathfrak{b}, \mathfrak{c}, \mathfrak{d}, \mathfrak{e}\}$ forms a packet.

We have accounted for the points on the interior, and the five points $\mathfrak{a}, \mathfrak{b}, \mathfrak{c}, \mathfrak{d}, \mathfrak{e}$. What remains are the five segments on the boundary. The two boundary segments $[\mathfrak{b}, \mathfrak{c}]$ and $[\mathfrak{e}, \mathfrak{d}]$ each contain the same number of accordions. We will form packets of order two by chosing one from each segment. We will combine two accordions that lie on the same vertical line. Concretely, this means that we group together the two patterns below, where $\varepsilon = (\varepsilon_1, \varepsilon_2)$ with $\delta_1 < \varepsilon_1 < \gamma_1$ and $s = \varepsilon_1 + \varepsilon_2$.

$\mathfrak{p}$	$\mathfrak{p}'$
$\mathfrak{p}_{(\mathfrak{d}.\mathfrak{e})}(\varepsilon) = \left\{ \begin{array}{cc} s & \alpha_2 \quad \varepsilon_2 \\ \alpha_1 \quad \varepsilon_1 \end{array} \right\}$	$\mathfrak{p}'_{(\mathfrak{d}.\mathfrak{e})}(\varepsilon) = \left\{ \begin{array}{cc} s \quad \varepsilon_1 \quad \alpha_1 \\ \varepsilon_2 \quad \alpha_2 \end{array} \right\}$
$\mathfrak{p}_{(\mathfrak{c},\mathfrak{b})}(\varepsilon) = \left\{ \begin{array}{cc} s \quad s \quad \varepsilon_2 \\ 0 \quad \varepsilon_1 \end{array} \right\}$	$\mathfrak{p}'_{(\mathfrak{c},\mathfrak{b})}(\varepsilon) = \left\{ \begin{array}{cc} s \quad \varepsilon_1 \quad 0 \\ \varepsilon_2 \quad s \end{array} \right\}$

We have the following values of $\Upsilon(\mathfrak{p})$.

$\mathfrak{p}$	$\mathcal{G}_\Gamma(\mathfrak{p})$	$\mathcal{G}_\Delta(\mathfrak{p}')$	$\Upsilon(\mathfrak{p})$
$\mathfrak{p}_{(\mathfrak{d}.\mathfrak{e})}(\varepsilon)$	$g(s)h_\alpha h_\varepsilon$	$h(s)h_\varepsilon g_\alpha$	$h_\varepsilon g(s)h(s)$
$\mathfrak{p}_{(\mathfrak{c},\mathfrak{b})}(\varepsilon)$	$q^s g(s)h_\varepsilon$	$h(s)^2 h_\varepsilon$	$-h_\varepsilon g(s)h(s)$

Therefore $\{\mathfrak{p}_{(\mathfrak{d},\mathfrak{e})}, \mathfrak{p}_{(\mathfrak{c},\mathfrak{b})}\}$ is a packet.

Finally, we find packets of order 3. The three segments from $\mathfrak{a}$ to $\mathfrak{b}$, from $\mathfrak{a}$ to $\mathfrak{e}$ and from $\mathfrak{c}$ to $\mathfrak{d}$ each have the same number of patterns, and these can be grouped together in packets of order 3. Let τ_1 be such that $\gamma_1 < \tau_1 < \beta_1$, and let $\theta_1 = \tau_1 - c_1 = \tau_1 - \gamma_1$. Determine τ_2 and θ_2 by $s = \tau_1 + \tau_2 = \theta_1 + \theta_2$.

$\mathfrak{p}$	$\mathfrak{p}'$
$\mathfrak{p}_{(\mathfrak{a}.\mathfrak{e})}(\tau) = \left\{ \begin{array}{cc} s & \alpha_2 \quad \tau_2 \\ \alpha_1 \quad \tau_1 \end{array} \right\}$	$\mathfrak{p}'_{(\mathfrak{a}.\mathfrak{e})}(\tau) = \left\{ \begin{array}{cc} s \quad \tau_1 \quad \alpha_1 \\ \tau_2 \quad \alpha_2 \end{array} \right\}$
$\mathfrak{p}_{(\mathfrak{a},\mathfrak{b})}(\tau) = \left\{ \begin{array}{cc} s \quad \theta_2 \quad \tau_2 \\ \theta_1 \quad \tau_1 \end{array} \right\}$	$\mathfrak{p}'_{(\mathfrak{a},\mathfrak{b})}(\tau) = \left\{ \begin{array}{cc} s \quad \tau_1 \quad \theta_1 \\ \tau_2 \quad \theta_2 \end{array} \right\}$
$\mathfrak{p}_{(\mathfrak{d},\mathfrak{c})}(\tau) = \left\{ \begin{array}{cc} s \quad \theta_2 \quad \delta_1 \\ \theta_1 \quad \delta_2 \end{array} \right\}$	$\mathfrak{p}'_{(\mathfrak{d},\mathfrak{c})}(\tau) = \left\{ \begin{array}{cc} s \quad \delta_2 \quad \theta_1 \\ \delta_1 \quad \theta_2 \end{array} \right\}$

Using Proposition 13.8 we find the following values for $\Upsilon(\mathfrak{p})$.

$\mathfrak{p}$	$\mathcal{G}_\Gamma(\mathfrak{p})$	$\mathcal{G}_\Delta(\mathfrak{p}')$	$\Upsilon(\mathfrak{p})$
$\mathfrak{p}_{(a,c)}(\tau)$	$g(s)h_\alpha h_\tau$	$h(s)g_\alpha h_\tau$	$h_\tau g(s)h(s)$
$\mathfrak{p}_{(a,b)}(\tau)$	$h(s)g_\theta h_\tau$	$h(s)h_\theta g_\tau$	$g(s)h(s)h_\theta - g(s)h(s)h_\tau$
$\mathfrak{p}_{(\partial,c)}(\tau)$	$h(s)h_\theta g_\delta$	$g(s)h_\theta h_\delta$	$-h_\theta g(s)h(s)$

We see that these three accordions form a packet. We have checked that the interior of the resotope consists of singleton packets, while the boundary of the resotope may be divided up into one packet of cardinality 5, and several packets of cardinalities 2 and 3.

If one looks at higher dimensional resotopes arising in larger rank examples, the above picture remains true: the interior of the resotope consists of singleton packets, and in any particular example one may, with some work, find a partitioning into packets. But the shapes of the resohedra, being truncated simplexes, are irregular, and there are always many cases to consider. Thus we turn to the more sophisticated approach outlined in Chapter 6.

Chapter Fourteen

Statement E Implies Statement D

We fix a nodal signature η. Let $B(\eta) = \{i | \eta_i = \Box\}$. Let $\mathcal{CP}_\eta(c_0, \cdots, c_d) \in 3_\Gamma$ be the following "cut and paste" virtual resotope

$$\mathcal{CP}_\eta(c_0, \cdots, c_d) = \sum_{T \subseteq B(\eta)} (-1)^{|T|} \mathcal{A}_s(c_0^T, \cdots, c_d^T), \tag{14.1}$$

where

$$c_i^T = \begin{cases} c_i & \text{if } i \in T, \\ \infty & \text{if } i \notin T. \end{cases}$$

We recall that the simplex $\mathrm{CP}_\eta(c_0, \cdots, c_d)$ is the set of Γ-accordions

$$\mathfrak{a} = \left\{ \begin{array}{cccc} s & \mu_1 & \cdots & \mu_d \\ \nu_1 & \cdots & \nu_d \end{array} \right\}$$

that satisfy the inequalities (6.22), with the convention that $\mu_0 = s$ and $\mu_{d+1} = 0$. Geometrically, this set is a simplex, and we will show that it is the support of $\mathcal{CP}_\eta(c_0, \cdots, c_d)$, though the latter virtual resotope is a superposition of resotopes whose supports include elements that are outside of $\mathrm{CP}_\eta(c_0, \cdots, c_d)$; it will be shown that the alternating sum causes such terms to cancel.

Finally, if $\mathfrak{a} \in \mathrm{CP}_\eta(c_0, \cdots, c_d)$ let $\theta(\mathfrak{a}, \eta)$ be the signature obtained from η by changing η_i to $*$ when the inequality

$$\mu_i - \mu_{i+1} \geqslant \begin{cases} c_i & \text{if } \eta_i = \Box, \\ 0 & \text{if } \eta_i = \circ, \end{cases}$$

is strict. Note that these are the inequalities defining $\mathfrak{a} \in \mathrm{CP}_\eta(c_0, \cdots, c_d)$. Strictly speaking $\mathfrak{a}$ and η do not quite determine $\theta(\mathfrak{a}, \eta)$ because it also depends on the c_i. We omit these data since they are fixed, while $\mathfrak{a}$ and η will vary.

PROPOSITION 14.1 *The support of $\mathcal{CP}_\eta(c_0, \cdots, c_d)$ is $\mathrm{CP}_\eta(c_0, \cdots, c_d)$. Suppose that $\mathfrak{a} \in \mathrm{CP}_\eta(c_0, \cdots, c_d)$. If τ is any signature, then the coefficient of $\mathfrak{a}_\tau$ in $\mathcal{CP}_\eta(c_0, \cdots, c_d)$ is zero unless τ is obtained from $\theta(\mathfrak{a}, \eta)$ by changing some $\Box$'s to $*$. If it is so obtained, the coefficient is $(-1)^\varepsilon$, where ε is the number of $\Box$'s in τ.*

Proof. Suppose the Γ-accordion $\mathfrak{a}$ does not satisfy (6.22). We will show that it does not appear in the support of $\mathcal{CP}_\eta(c_0, \cdots, c_d)$. By assumption $\mu_i - \mu_{i+1} < c_i$ for some $i \in B(\eta)$. We group the subsets of $B(\eta)$ into pairs T, T' where $T = T' \cup \{i\}$. It is clear that $\mathfrak{a}$ occurs in $\mathcal{A}_s(c_0^T, \cdots, c_d^T)$ if and only if it occurs in $\mathcal{A}_s(c_0^{T'}, \cdots, c_d^{T'})$, and with the same signature. Since these have opposite signs, their contributions cancel. This proves that the support of $\mathcal{CP}_\eta(c_0, \cdots, c_d)$ is contained in the simplex $\mathrm{CP}_\eta(c_0, \cdots, c_d)$. The opposite inclusion will be clear from the precise description of the coefficients, which we turn to next.

We note that $\theta(\mathfrak{a}, \eta) = \theta_0 \cdots \theta_d$ where

$$\theta_i = \begin{cases} \square & \text{if } \mu_i - \mu_{i+1} = c_i, \\ \circ & \text{if } \mu_i - \mu_{i+1} = 0, \\ * & \text{otherwise.} \end{cases}$$

We emphasize that if $\theta_i = \square$ then $i \in B(\eta)$, while if $\theta_i = \circ$ then $i \notin B(\eta)$. (The case $\theta_i = *$ can arise whether or not $i \in B(\eta)$.)

Suppose that $\mathfrak{a} \in \mathrm{CP}_\eta(c_0, \cdots, c_d)$. In order for $\mathfrak{a}_\tau$ to have a nonzero coefficient, it must appear as the coefficient of $\mathfrak{a}$ in $A_s(c_0^T, \cdots, c_d^T)$ for some subset T of $B(\eta)$. We will prove that if τ is the signature of $\mathfrak{a}$ in this resotope we have

$$\tau_i = \begin{cases} \circ & \text{if } \mu_i - \mu_{i+1} = 0, \text{ in which case } i \notin B(\eta); \\ \square & \text{if } i \in T; \\ * & \text{otherwise.} \end{cases} \tag{14.2}$$

First, if $i \in T$ then $c_i^T = c_i$ so $\mu_i - \mu_{i+1} \leqslant c_i$, for we have already stipulated (by assuming $\mathfrak{a} \in \mathrm{CP}_\eta(c_0, \cdots, c_d)$) that $\mu_i - \mu_{i+1} \geqslant c_i$. Therefore $\mu_i - \mu_{i+1} = c_i$ when $i \in T$, and so $\tau_i = \square$ when $i \in T$. And if $i \notin T$, the signature of τ is definitely not $\square$ since $c_i^T = \infty$; if $i \in B(\eta) - T$ it also cannot be $\circ$ since $\mu_i - \mu_{i+1} \geqslant c_i > 0$. This proves (14.2).

It is clear from (14.2) that τ is obtained from $\theta(\mathfrak{a})$ by changing some $\square$'s to $*$'s, and which ones are changed determines T. This point is important since it shows that (unlike the case where $\mathfrak{a} \notin \mathrm{CP}_\eta(c_0, \cdots, c_d)$) a given $\mathfrak{a}_\tau$ can only appear in only one term in (14.1), so there cannot be any cancellation. If τ is obtained from $\theta(\mathfrak{a})$ by changing some $\square$'s to $*$'s then it does appear in $A_s(c_0^T, \cdots, c_d^T)$ for a unique T and so $\mathfrak{a}_\tau$ appears in $\mathcal{CP}_\eta(c_0, \cdots, c_d)$ with a nonzero coefficient. The sign with which it appears is $(-1)^{|T|}$, and T we have noted is the set of i for which $\tau_i = \square$. $\square$

THEOREM 14.2 *Statement E implies Statement D.*

Proof. Let $\mathfrak{a} \in \mathrm{CP}_\eta(c_0, \cdots, c_d)$ and let $\sigma = \theta(\mathfrak{a}, \eta)$. What we must show is that (6.23) implies (6.20). We extend the function G_Γ from the set of decorated Γ-accordions to the free abelian group 3_Γ by linearity. Also the involution $\mathfrak{a}_\eta \longmapsto \mathfrak{a}'_\eta$ on decorated accordions induces an isomorphism $3_\Gamma \longrightarrow 3_\Delta$ that we will denote $\mathcal{A} \longmapsto \mathcal{A}'$.

Then (6.20) can be written $\mathcal{G}_\Gamma(\mathcal{A}) = \mathcal{G}_\Delta(\mathcal{A}')$. By the principle of inclusion-exclusion (Stanley [68], page 64), we have

$$A_s(c_0, \cdots, c_d) = \sum_{T \subseteq B(\eta)} \mathcal{CP}_{\eta^T}(c_0^T, \cdots, c_d^T),$$

where if T is a subset of $B(\eta)$ then η^T is the signature obtained by changing η_i from $\square$ to $\circ$ for all $i \in T$. This means that if we show $\mathcal{G}_\Gamma(\mathcal{C}) = \mathcal{G}_\Delta(\mathcal{C}')$ when $\mathcal{C} = \mathcal{CP}_\eta(c_0, \cdots, c_d)$ then (6.20) will follow. The left-hand side in this identity is a sum of $\mathcal{G}_\Gamma(\mathfrak{a}_\tau)$ with $\mathfrak{a}$ in $\mathrm{CP}_\eta(c_0, \cdots, c_d)$, and the coefficient of $\mathfrak{a}_\tau$ in this sum is the same as its coefficient in $\Lambda_\Gamma(\mathfrak{a}, \sigma)$ by Proposition 14.1. $\square$

Chapter Fifteen

Evaluation of Λ_Γ and Λ_Δ, and Statement G

Let η be a nodal signature, and let σ be a subsignature. Let

$$\mathfrak{a} = \left\{ \begin{matrix} s & \alpha_1 & \alpha_2 & \cdots & \alpha_d \\ \beta_1 & \beta_2 & \cdots & \beta_d \end{matrix} \right\}$$

be an accordion belonging to the open facet $\mathcal{S}_\sigma$ of $\mathrm{CP}_\eta(c_0, \cdots, c_d)$. Assuming that $n|s$ we will evaluate $\Lambda_\Gamma(\mathfrak{a}, \sigma)$.

We will denote

$$V(a, b) = (q-1)^a q^{(d+1)s-b}, \qquad V(a) = V(a, a).$$

Let

$$\varepsilon_\Gamma(\sigma) = \varepsilon_\Gamma = \left\{ \begin{matrix} 1 & \text{if } \sigma_0 = \square, \\ 0 & \text{otherwise,} \end{matrix} \right.$$

$$\mathcal{K}_\Gamma(\sigma) = \mathcal{K}_\Gamma = \{i|1 \leqslant i \leqslant d, \sigma_i = \square, \sigma_{i-1} \neq \circ \}, \qquad k_\Gamma = |\mathcal{K}_\Gamma|,$$
$$\mathcal{N}_\Gamma(\sigma) = \mathcal{N}_\Gamma = \{i|1 \leqslant i \leqslant d, \sigma_i = \square, \sigma_{i-1} = \circ \}, \qquad n_\Gamma = |\mathcal{N}_\Gamma|,$$

and

$$\mathcal{C}_\Gamma(\sigma) = \mathcal{C}_\Gamma =$$
$$\{i|1 \leqslant i \leqslant d, \sigma_0, \sigma_1, \cdots, \sigma_{i-1} \text{ not all } \circ \text{ and either } i \in \mathcal{N}_\Gamma \text{ or } \sigma_i = *\}.$$

$$(15.1)$$

Let $c_\Gamma = |\mathcal{C}_\Gamma|$, and let t_Γ be the number of i with $1 \leqslant i \leqslant d$ and $\sigma_i \neq *$. Given a set of indices $\Sigma = \{i_1, \cdots, i_k\} \subseteq \{1, 2, \cdots, d\}$, let

$$\delta_n(i_1, \cdots, i_k) = \delta_n(\Sigma) = \delta_n(\Sigma; \mathfrak{a}) = \left\{ \begin{matrix} 1 & \text{if } n \text{ divides } \alpha_{i_1}, \cdots, \alpha_{i_k}, \\ 0 & \text{otherwise.} \end{matrix} \right. \quad (15.2)$$

Let

$$\chi_\Gamma(\mathfrak{a}, \sigma) = \chi_\Gamma = \prod_{i \in \mathcal{C}_\Gamma(\sigma)} \delta_n(i).$$

Finally, let

$$a_\Gamma(\sigma) = a_\Gamma = 2(d - t_\Gamma + n_\Gamma) + \left\{ \begin{matrix} -1 & \text{if } \sigma_0 = \circ \\ 0 & \text{if } \sigma_0 = \square \\ 1 & \text{if } \sigma_0 = * \end{matrix} \right\} + \left\{ \begin{matrix} 1 & \text{if } \sigma_d = \circ \\ 0 & \text{if } \sigma_d \neq \circ \end{matrix} \right\}.$$

PROPOSITION 15.1 *Assume that $n|s$. Given a Γ-accordion*

$$\mathfrak{a} = \left\{ \begin{matrix} s & \alpha_1 & \alpha_2 & \cdots & \alpha_d \\ \beta_1 & \beta_2 & \cdots & \beta_d \end{matrix} \right\}$$

and an associated signature $\sigma \subseteq \sigma_{\mathfrak{I}}$ not containing the sequence $\bigcirc, \square$, then

$$\mathcal{G}_\Gamma(\mathfrak{a}, \sigma) = (-1)^{\varepsilon_\Gamma} \chi_\Gamma \cdot V(a_\Gamma, a_\Gamma + d_\Gamma), \tag{15.3}$$

where

$$d_\Gamma = \left(\sum_{\substack{1 \leqslant i \leqslant d \\ \sigma_i = \square}} (1 + \delta_n(i)) \right) + \left\{ \begin{array}{ll} 1 & \text{if } \sigma_0 = \square \\ 0 & \text{if } \sigma_0 \neq \square \end{array} \right\}.$$

Recall that any subsignature σ containing the string $\bigcirc \square$ has $\mathcal{G}_\Gamma(\mathfrak{a}, \sigma) = 0$. We will abuse notation and rewrite the definition of $\mathcal{G}_\Gamma(\mathfrak{a}, \sigma)$ as

$$\mathcal{G}_\Gamma(\mathfrak{a}, \sigma) = \mathcal{G}_\Gamma(\mathfrak{a}_\sigma) = \prod_{x \in \mathfrak{a}} f_\sigma(x),$$

where

$$f_\sigma(x) = \left\{ \begin{array}{ll} g(x) & \text{if } x \text{ is boxed in } \mathfrak{a}_\sigma \text{ but not circled,} \\ q^x & \text{if } x \text{ is circled but not boxed,} \\ h(x) & \text{if } x \text{ is neither boxed nor circled,} \\ 0 & \text{if } x \text{ is both boxed and circled.} \end{array} \right.$$

This is an abuse of notation, since f_σ is not a function; it depends not only on the numerical value x but also its location in the decorated accordion $\mathfrak{a}_\sigma$. However this should cause no confusion.

Proof. Using the signature σ to determine the rules for boxing and circling in $\mathfrak{a}$ we see that if $\sigma_0 = \square$, then $f_\sigma(s) = g(s) = (-1) \cdot q^{s-1}$. The (-1) here accounts for the $(-1)^{\varepsilon_\Gamma}$ in (15.3). If $\sigma_i = \square$ for $i > 0$, then by Proposition 8.1

$$f_\sigma(\alpha_i) f_\sigma(\beta_i) = g(\alpha_i) g(\beta_i) = \left\{ \begin{array}{ll} q^{s-1} & \text{if } n \nmid \alpha_i, \\ q^{s-2} & \text{if } n | \alpha_i. \end{array} \right.$$

If $\sigma_0 = \bigcirc$, then $s = \alpha_1$, $\beta_1 = 0$, and $f_\sigma(s) = q^s$. If $\sigma_i = \bigcirc$, $0 < i < d$, then $\alpha_i = \alpha_{i+1}$ and $\beta_i = \beta_{i+1}$ so that while the circling in the accordion strictly speaking occurs at α_i and β_{i+1}, we may equivalently consider it to occur at α_i and β_i for bookkeeping purposes and

$$f_\sigma(\alpha_i) f_\sigma(\beta_{i+1}) = f_\sigma(\alpha_i) f_\sigma(\beta_i) = q^s.$$

And if $\sigma_d = \bigcirc$, then $\alpha_d = 0$ and $\beta_d = s$, so that

$$f_\sigma(\alpha_d) f_\sigma(\beta_d) = h(s) = (q-1)q^{s-1}.$$

Finally if $\sigma_0 = *$, then $f_\sigma(s) = (q-1)q^{s-1}$. If $\sigma_i = *, 1 \leqslant i \leqslant d$ then

$$f_\sigma(\alpha_i) f_\sigma(\beta_i) = h(\alpha_i) h(\beta_i) = \left\{ \begin{array}{ll} (q-1)^2 q^{s-2} & \text{if } n | \alpha_i, \\ 0 & \text{if } n \nmid \alpha_i. \end{array} \right.$$

Now note that the assumption that σ does not contain the string $\bigcirc, \square$ implies that $n_\Gamma = 0$, simplifying the definitions of χ_Γ and a_Γ above. The case of $\sigma_i = \square$ is seen to account for the d_Γ defined above, the $\sigma_i = *$ account for both the χ_Γ and the a_Γ. However, one does need to count somewhat carefully at the ends of the accordion according to the above cases. In particular, we see that $\sigma_0 = \circ$ implies $\alpha_1 = s$, so that if j is the first index with $\sigma_j \neq \circ$, then $\sigma_j = *$ by assumption. But then $\alpha_j = s$ and the divisibility condition $n|\alpha_j$ is automatic, hence redundant and omitted from the definition. □

LEMMA 15.2 *Given a signature σ that does not contain the sequence $\circ\square$ and $d_\Gamma(\sigma)$ as defined in Proposition 15.1, we may write $d_\Gamma = k_\Gamma + \varepsilon_\Gamma + \sum_{i \in \mathcal{K}_\Gamma(\sigma)} \delta_n(i)$. Then for any m with $0 \leqslant m < d_\Gamma$*

$$\binom{d_\Gamma}{m} = k_\Gamma + \binom{\varepsilon_\Gamma}{m} + \sum_{i \in \mathcal{K}_\Gamma(\sigma)} \delta_n(i) \binom{k_\Gamma + \varepsilon_\Gamma}{m-1} +$$

$$\cdots + \sum_{\{i_1,\ldots,i_l\} \subseteq \mathcal{K}_\Gamma(\sigma)} \delta_n(i_1,\ldots,i_l) \binom{k_\Gamma + \varepsilon_\Gamma}{m-l} + \cdots$$

where we understand each of the binomial coefficients to be 0 if the lower entry is either negative or larger than the upper entry.

Proof. The result follows from repeated application of the identity

$$\binom{c + \delta_n(i)}{m} = \binom{c}{m} + \delta_n(i) \binom{c}{m-1},$$

valid for any constants c and m and index i, While d_Γ contains divisibility conditions and hence depends on $\mathfrak{a}$, $k_\Gamma + \varepsilon_\Gamma$ is an absolute constant depending only on the signature σ. □

THEOREM 15.3 *Fix a nodal signature η, and assume that $n|s$. Given an accordion $\mathfrak{a} \in S_\sigma$ with subsignature $\sigma = (\sigma_0, \sigma_1, \ldots, \sigma_d) \subset \eta$,*

$$\Lambda_\Gamma(\mathfrak{a}, \sigma) = (-1)^{n_\Gamma + \varepsilon_\Gamma} \chi_\Gamma \sum_{x=0}^{k_\Gamma} \sum_{\substack{\Sigma \subseteq \mathcal{K}_\Gamma(\sigma) \\ |\Sigma|=x}} \delta_n(\Sigma)(-1)^x V(a_\Gamma + x, a_\Gamma + k_\Gamma), \quad (15.4)$$

where the inner sum ranges over all possible subsets of cardinality x in $\mathcal{K}_\Gamma$.

Before giving the proof, let us do an example. Let

$$\sigma = (\square, *, \circ, \square, *, \square, *) \subseteq \eta = (\square, \circ, \circ, \square, \square, \square, \circ)$$

then one can read off the following data from the signature:

$$n_\Gamma = 1, \quad \varepsilon_\Gamma = 1, \quad \chi_\Gamma = \delta_n(1, 3, 4, 6),$$

$$t_\Gamma = 3, \quad \mathcal{K}_\Gamma = \{5\}, \quad k_\Gamma = 1, \quad a_\Gamma = 8$$

so

$$\Lambda_\Gamma(\sigma) = \delta_n(1, 3, 4, 6)\,(V(8, 9) - \delta_n(5)V(9, 9))$$

Proof. We may express

$$V(a, a+b) = \sum_{u=0}^{b} (-1)^u \binom{b}{u} V(a+u, a+u) \tag{15.5}$$

from the binomial theorem, given the definition $V(a,b) = (q-1)^a q^{(d+1)s-b}$.

From the definition we have

$$\Lambda_\Gamma(\mathfrak{a}, \sigma) = \mathcal{G}_\Gamma(\mathfrak{a}, \sigma) - \sum_{\sigma^{(1)}} \mathcal{G}_\Gamma(\mathfrak{a}, \sigma^{(1)}) + \ldots + (-1)^i \sum_{\sigma^{(i)}} \mathcal{G}_\Gamma(\mathfrak{a}, \sigma^{(i)}) + \ldots,$$

where the sums run over $\sigma^{(i)} \subseteq \sigma$ obtained from σ by replacing exactly i occurrences of $\square$ by $*$. We will apply Proposition 15.1 to evaluate these terms, then simplify.

If the sequence $\circ\square$ appears within σ, then we call signature Γ-*non-strict*, as any corresponding short pattern $\mathfrak{T}$ with $\sigma \subseteq \sigma_\mathfrak{T}$ is non-strict, and by definition $\mathcal{G}_\Gamma(\mathfrak{T}, \sigma) = 0$. Thus the alternating sum for Λ_Γ will only contain nonzero contributions from subsignatures when all such $\square$'s occurring as part of a $\circ\square$ string in σ have been removed (i.e., changed to an $*$). Upon doing this, the signature will no longer possess any subwords of the form $\circ\square$, and we may again apply the above formula for $\mathcal{G}_\Gamma$ to these subsignatures. This is reflected in the definition of a_Γ and in the statement of the Theorem.

We first assume that $\sigma_0 \neq \square$ and that $\circ\square$ does not occur within σ. Then

$$G_\Gamma(\mathfrak{a}, \sigma) = \chi_\Gamma V(a_\Gamma, a_\Gamma + d_\Gamma) = \chi_\Gamma \sum_{u=0}^{d_\Gamma} (-1)^u \binom{d_\Gamma}{u} V(a_\Gamma + u)$$

$$= \chi_\Gamma \sum_{u=0}^{d_\Gamma} (-1)^u V(a_\Gamma + u) \sum_{l=0}^{k_\Gamma} \sum_{\{i_1, \ldots, i_l\} : \sigma_{i_j} = \square} \binom{k_\Gamma}{u-l} \delta_n(i_1, \ldots, i_l)$$

$$= \chi_\Gamma \sum_{l=0}^{k_\Gamma} \sum_{\{i_1, \ldots, i_l\} : \sigma_{i_j} = \square} \delta_n(i_1, \ldots, i_l) \sum_{u=l}^{d_\Gamma} (-1)^u \binom{k_\Gamma}{u-l} V(a_\Gamma + u)$$

where we have used Proposition 15.1 and Lemma 15.2 in the first two steps, and in the last step have simply interchanged the order of summation.

By similar calculation (still assuming that $\sigma_0 \neq \square$ for simplicity of exposition) we have

$$\sum_{\sigma^{(m)} \subseteq \sigma} G_\Gamma(\mathfrak{a}, \sigma^{(m)})$$

$$= \chi_\Gamma \sum_{\{i_1, \ldots, i_m\} : \sigma_{i_j} = \square} \delta_n(i_1, \ldots, i_m) \sum_{l=0}^{k_\Gamma - m} \sum_{\{i'_1, \ldots, i'_l\} : \sigma_{i'_{j'}} \neq \sigma_{i_j}} \delta_n(i_1, \ldots, i_l)$$

$$\times \sum_{u=l}^{d_\Gamma - 2m} (-1)^u \binom{k_\Gamma - m}{u-l} V(a_\Gamma + 2m + u)$$

where we can write the upper bound on the sum over u as an absolute constant, since either the divisibility conditions are satisfied and the upper bound (equal to $d_\Gamma(\sigma^{(m)})$) is indeed $d_\Gamma(\sigma) - 2m$ or else the term is 0. Simplifying by combining the two sums with divisibility conditions, we have

$$\sum_{\sigma^{(m)} \subseteq \sigma} G_\Gamma(\mathfrak{a}, \sigma^{(m)}) = \chi_\Gamma \sum_{l=0}^{k_\Gamma - m} \sum_{\{i_1, \ldots, i_{m+l}\}: \sigma_{i_j} = \square} \binom{m+l}{m} \delta_n(i_1, \ldots, i_{m+l})$$

$$(-1)^l \sum_{v=0}^{d_\Gamma - 2m - l} (-1)^v \binom{k_\Gamma - m}{v} V(a_\Gamma + 2m + l + v).$$

Hence

$$\Lambda_\Gamma(\mathfrak{a}, \sigma) = \chi_\Gamma \sum_{m=0}^{k_\Gamma} (-1)^m \sum_{l=0}^{k_\Gamma - m} \sum_{\{i_1, \ldots, i_{m+l}\}: \sigma_{i_j} = \square} \binom{m+l}{m} \delta_n(i_1, \ldots, i_{m+l})$$

$$(-1)^l \sum_{v=0}^{d_\Gamma - 2m - l} (-1)^v \binom{k_\Gamma - m}{v} V(a_\Gamma + 2m + l + v)$$

$$= \chi_\Gamma \sum_{m=0}^{k_\Gamma} \sum_{x=m}^{k_\Gamma} \sum_{\{i_1, \ldots, i_x\}: \sigma_{i_j} = \square} \binom{x}{m} \delta_n(i_1, \ldots, i_x)(-1)^x$$

$$\sum_{v=0}^{d_\Gamma - x - m} (-1)^v \binom{k_\Gamma - m}{v} V(a_\Gamma + m + x + v)$$

$$= \chi_\Gamma \sum_{x=0}^{k_\Gamma} \sum_{\{i_1, \ldots, i_x\}: \sigma_{i_j} = \square} \delta_n(i_1, \ldots, i_x)(-1)^x$$

$$\sum_{m=0}^{x} \binom{x}{m} \sum_{v=0}^{d_\Gamma - x - m} (-1)^v \binom{k_\Gamma - m}{v} V(a_\Gamma + m + x + v) \qquad (15.6)$$

where in the first step we changed the sum over l to a sum over $x = m + l$ and interchanged the order of summation in the second step. Now let $w = m + v$, so (15.6) equals

$$\sum_{m=0}^{x} (-1)^m \sum_{w=m}^{d_\Gamma - x} (-1)^w \binom{x}{m} \binom{k_\Gamma - m}{w - m} V(a_\Gamma + x + w) =$$

$$\sum_{w=0}^{d_\Gamma - x} (-1)^w V(a_\Gamma + x + w) \sum_{m=0}^{w} (-1)^m \binom{x}{m} \binom{k_\Gamma - m}{w - m}. \qquad (15.7)$$

But

$$\sum_{m=0}^{w} (-1)^m \binom{x}{m} \binom{k_\Gamma - m}{w - m} = \binom{k_\Gamma - x}{w}, \qquad (15.8)$$

so combining (15.6) and (15.7) and applying (15.8)

$$\Lambda_\Gamma(\mathfrak{a}, \sigma) =$$

$$\chi_\Gamma \sum_{m=0}^{k_\Gamma} (-1)^m \sum_{\{i_1,\dots,i_{m+l}\}:\sigma_{i_j}=\square} \delta_n(i_1, \cdots, i_x)$$

$$(-1)^x \sum_{w=0}^{d_\Gamma - x} (-1)^w \binom{k_\Gamma - x}{w} V(a_\Gamma + x + w)$$

The cases where $\sigma_0 = \square$ or where $\circ\,\square$ appears in the signature follow by a straightforward generalization. $\qquad\qquad\square$

We turn now to the evaluation of $\Lambda_\Delta(\mathfrak{a}', \sigma)$, where σ is unchanged and

$$\mathfrak{a}' = \left\{ \begin{array}{cccccc} \beta_1 & & \beta_2 & \cdots & \beta_d & s \\ & \alpha_1 & & \alpha_2 & \cdots & \alpha_d \end{array} \right\}.$$

Let

$$\varepsilon_\Delta(\sigma) = \varepsilon_\Delta = \left\{ \begin{array}{ll} 1 & \text{if } \sigma_d = \square, \\ 0 & \text{otherwise}, \end{array} \right.$$

$$\mathcal{K}_\Delta = \{0 < i \leqslant d \mid \sigma_{i-1} = \square, \sigma_i \neq \circ\}, \quad k_\Delta(\sigma) = k_\Delta = |\mathcal{K}_\Delta|, \quad (15.9)$$

$$\mathcal{N}_\Delta = \{0 < i \leqslant d \mid \sigma_{i-1} = \square, \sigma_i = \circ\}, \quad n_\Delta(\sigma) = n_\Delta = |\mathcal{N}_\Delta|,$$

$$\mathcal{C}_\Delta = \{1 \leqslant i \leqslant d \mid \sigma_i, \sigma_{i+1}, \cdots, \sigma_d \text{ not all } \circ \text{ and either } \sigma_{i-1} = * \text{ or } i \in \mathcal{N}_\Delta\},$$

$$(15.10)$$

$$\chi_\Delta(\mathfrak{a}, \sigma) = \chi_\Delta = \prod_{i \in \mathcal{C}_\Delta(\sigma)} \delta_n(i), \qquad t_\Delta = |\{0 \leqslant i < d \mid \sigma_i = \eta_i\}|$$

$$a_\Delta(\sigma) = a_\Delta = 2(d - (t_\Delta - n_\Delta)) + \left\{ \begin{array}{ll} -1 & \text{if } \sigma_d = \circ \\ 0 & \text{if } \sigma_d = \square \\ 1 & \text{if } \sigma_d = * \end{array} \right\} + \left\{ \begin{array}{ll} 1 & \text{if } \sigma_0 = \circ \\ 0 & \text{if } \sigma_0 \neq \circ \end{array} \right\}.$$

We give $\delta_n(i_1, \cdots, i_k)$ the same meaning as before: it is $\delta_n(i_1, \cdots, i_k; \mathfrak{a})$. But since the top row of $\mathfrak{a}'$ is given in terms of the β's, note that it can also be described as 1 if n divides $\beta_{i_1}, \cdots, \beta_{i_k}$ and 0 otherwise. Indeed, $n|s$ so $n|\alpha_i$ if and only if $n|\beta_i = s - \alpha_i$.

THEOREM 15.4 *With notation as above we have*

$$\Lambda_\Delta(\mathfrak{a}', \sigma) =$$

$$(-1)^{n_\Delta + \varepsilon_\Delta} \chi_\Delta \sum_{x=0}^{k_\Delta} \sum_{\substack{\Sigma \subseteq \mathcal{K}_\Delta \\ |\Sigma| = x}} \delta_n(\Sigma)(-1)^x V(a_\Delta + x, a_\Delta + k_\Delta).$$

where (as defined above) the inner sums range over subsets of $\mathcal{K}_\Delta$.

Proof. We can apply our previous work by noting that

$$\Lambda_\Delta(\mathfrak{a}', \sigma) = \Lambda_\Gamma(\widetilde{\mathfrak{a}}, \widetilde{\sigma}),$$

where

$$\tilde{a} = \left\{ \begin{matrix} s & & \beta_d & & \beta_{d-1} & & \cdots & & \beta_1 \\ & \alpha_d & & \alpha_{d-1} & & \cdots & & \alpha_1 & \end{matrix} \right\},$$

and $\tilde{\sigma} = \sigma_d \sigma_{d-1} \cdots \sigma_0$. Roughly speaking we can just take the mirror image of our previous formula. But there is one point of caution: in going from a to $\tilde{a}$ we reflected σ in the range 0 to d, while we reflected α in the range 1 to d (and changed it to β, which has no effect on δ). This means the $\mathcal{C}_\Delta(\sigma)$, if it is to be the set of locations where the congruences are taken in evaluating δ, is *not* the mirror image of $\mathcal{C}_\Gamma(\tilde{a})$ in the range 0 to d, but the shift of that mirror image to the right by 1, which makes $\mathcal{C}_\Delta(\sigma)$, like $\mathcal{C}_\Gamma(\sigma)$, a subset of the range from 1 to d. There are corresponding adjustments in the definitions of $\mathcal{K}_\Delta$ and $\mathcal{N}_\Delta$. □

Let Π be an f-packet (as defined in Chapter 6 before Statement F) for the nodal signature η. We recall that the packet Π in Statement F intersects each open f-facet S_σ in a unique element a, and so a is determined by σ. Let $\mathcal{A}_\Gamma(\sigma)$ denote the set of Γ-admissible sets for σ, and let $\mathcal{A}_\Delta(\sigma)$ denote the set of Δ-admissible sets. We may reformulate Statement F in the following way.

Statement G. *Given a nodal signature η and a corresponding f-packet Π,*

$$\sum_\sigma (-1)^{n_\Gamma(\sigma)+\varepsilon_\Gamma(\sigma)} \sum_{\substack{0 \leqslant x \leqslant k_\Gamma(\sigma) \\ \Sigma \in \mathcal{A}_\Gamma(\sigma) \\ |\Sigma - \mathcal{C}_\Gamma(\sigma)|=x}} (-1)^x V(a_\Gamma + x, a_\Gamma + k_\Gamma) \, \delta_n(\Sigma, a) =$$

$$\sum_\sigma (-1)^{n_\Delta(\sigma)+\varepsilon_\Delta(\sigma)} \sum_{\substack{0 \leqslant x \leqslant k_\Delta(\sigma) \\ \Sigma \in \mathcal{A}_\Delta(\sigma) \\ |\Sigma - \mathcal{C}_\Delta(\sigma)|=x}} (-1)^x V(a_\Delta + x, a_\Delta + k_\Delta) \, \delta_n(\Sigma, a),$$

$$(15.11)$$

where the outer sums are over f-subsignatures σ of η.

We have restored a to the notation $\delta_n(\Sigma; a)$ from which it was suppressed in Theorems 15.3 and 15.4, because the dependence of these terms on a – or, equivalently, on σ – will now become our most important issue.

THEOREM 15.5 *Statement G implies Statement F.*

Proof. This follows immediately from Theorems 15.3 and 15.4, since in (6.24) each such f-subsignature σ appears exactly once on each side. □

Chapter Sixteen

Concurrence

This chapter contains purely combinatorial results that are needed for the proof of Statement G. The motivation for these results comes from the appearance of divisibility conditions through the factor $\delta_n(\Sigma; \mathfrak{a})$ defined in (15.2) that appears in Theorems 15.3 and 15.4. We refer to the discussion of Statement G in Chapter 6 for an informal discussion of the results of this chapter.

Let $0 \leqslant f \leqslant d$. In Chapter 6 we defined bijections $\phi_{\sigma,\tau} : \mathcal{S}_\sigma \longrightarrow \mathcal{S}_\tau$ between the open f-facets, and a related equivalence relation, whose classes we call f-packets. According to Statement F, the sum of $\Lambda_\Gamma(\mathfrak{a}, \sigma)$ over an f-packet is equal to the corresponding sum of $\Lambda_\Delta(\mathfrak{a}', \sigma)$. Moreover in Theorems 15.3 and 15.4, we have rewritten Λ_Γ and Λ_Δ as sums over ordered subsets of $\mathcal{K}_\Gamma$ and $\mathcal{K}_\Delta$. In order to prove Statement F, we will proceed by identifying terms in the resulting double sum that can be matched, and that is the aim of the results of this chapter.

DEFINITION 16.1 (Concurrence) *Let σ and τ be subsignatures of η that have the same number of $*$'s. Fix two subsets $\Sigma = \{j_1, \cdots, j_l\}$ and $\Sigma' = \{j_1', \ldots, j_l'\}$ of $\{1, 2, \cdots, d\}$ of equal cardinality, and arranged in ascending order:*

$$0 \leqslant j_1 < j_2 < \cdots < j_l \leqslant d, \qquad 0 \leqslant j_1' < j_2' < \cdots < j_l' \leqslant d.$$

We say that the pairs (σ, Σ) and (τ, Σ') concur if the following conditions are satisfied. We require that for $1 \leqslant m \leqslant l$ the two sets

$$\{t \mid j_m \leqslant t \leqslant d, \sigma_t = *\}, \qquad \{t \mid j_m' \leqslant t \leqslant d, \tau_t = *\} \tag{16.1}$$

have the same cardinality, and that $\eta_i = \circ$ for

$$\min(j_m, j_m') \leqslant i < \max(j_m, j_m'). \tag{16.2}$$

Concurrence is an equivalence relation. To give an example, let

$$\eta = (\eta_0, \eta_1, \ldots, \eta_5) = (\circ, \square, \circ, \circ, \square, \square).$$

The pairs

$$((*, \square, *, \circ, *, \square), \{2, 4, 5\}); \qquad ((\circ, *, *, *, \square, \square), \{2, 3, 5\})$$

concur. However

$$((*, \square, *, \circ, *, \square), \{2, 4, 5\}); \qquad ((\circ, *, *, *, \square, \square), \{2, 4, 5\})$$

do not, as the number of $*$'s to the right of σ_4, τ_4 differ.

PROPOSITION 16.2 *Suppose that the pairs (σ, Σ) and (τ, Σ') concur. If $\phi_{\sigma,\tau}(\mathfrak{a}) = \mathfrak{b}$, then we have $\alpha_{j_m} = \mu_{j_m'}$ $(1 \leqslant m \leqslant l)$, where*

$$\mathfrak{a} = \left\{ \begin{matrix} s & \alpha_1 & \cdots & \alpha_d \\ \beta_1 & \cdots & \beta_d \end{matrix} \right\}, \quad \mathfrak{b} = \left\{ \begin{matrix} s & \mu_1 & \cdots & \mu_d \\ \nu_1 & \cdots & \nu_d \end{matrix} \right\}.$$

This implies that

$$\delta_n(\Sigma; \mathfrak{a}) = \delta_n(\Sigma'; \mathfrak{b}),$$

which can be used to compare the contributions of these ordered subsets to $\Lambda_\Gamma(\mathfrak{a}, \sigma)$ and $\Lambda_\Delta(\mathfrak{a}', \sigma)$ with the corresponding contributions to $\Lambda_\Gamma(\mathfrak{b}, \tau)$ and $\Lambda_\Delta(\mathfrak{b}', \tau)$ in the formulas of Theorems 15.3 and 15.4.

Proof. It is sufficient to check this when $\mathfrak{a}$ is a vertex of $\overline{S_\sigma}$. Indeed, both α_{j_m} and $\mu_{j'_m}$ are affine-linear functions of $\alpha_1, \cdots, \alpha_d$, so if they are the same when $\mathfrak{a} = \mathfrak{a}_k$ is a vertex, they will be the same for convex combinations of the vertices, that is, for all elements of $\overline{S_\sigma}$. Because $\mathfrak{a}_k$ is a vertex of $\overline{S_\sigma}$, $\sigma_k = *$; if σ_k is the r-th $*$ in σ, then by definition $\phi_{\sigma,\tau}(\mathfrak{a}) = \mathfrak{a}_l$ where τ_l is the r-th $*$ in τ. This is a consequence of the definition of $\phi_{\sigma,\tau}$. Now our assumption on the cardinality of the two sets (16.1) implies that $k \leqslant j_m$ if and only if $l \leqslant j'_m$.

Now we prove that $\alpha_{j_m} = \mu_{j'_m}$. There are two cases, depending on whether $j_m \leqslant k$ (and so $j'_m \leqslant l$) or not. First suppose that $j_m \leqslant k$ and $j'_m \leqslant l$. Then we have $\alpha_i - \alpha_{i+1} = c'_i$ for all i except k and $\mu_i = \mu_{i+1} = c'_i$ for all i except l, and $\alpha_0 = s = \mu_0$. This means that $\alpha_i = \mu_i$ when $i \leqslant \min(k, l)$, *a fortiori* when $i \leqslant \min(j_m, j'_m)$. Suppose for definiteness that $j_m \leqslant j'_m$, so $\min(j_m, j'_m) = j_m$. Thus we have proved that $\alpha_{j_m} = \mu_{j_m}$. Since by hypothesis $\eta_{j_m} = \eta_{j_m+1} = \cdots = \eta_{j'_m-1} = \bigcirc$ we also have $\mu_{j_m} = \mu_{j_m+1} = \cdots \mu_{j'_m}$ and therefore $\alpha_{j_m} = \mu_{j'_m}$. The case where $j_m \geqslant j'_m$ is similar, and the case where $j_m \leqslant k$ and $j'_m \leqslant l$ is settled.

Next suppose that $j_m > k$ and so $j'_m > l$. Then $\alpha_i - \alpha_{i+1} = c'_i$ for all i except k and $\mu_i = \mu_{i+1} = c'_i$ for all i except l, and $\alpha_{d+1} = 0 = \mu_{d+1}$, and we get $\alpha_i = \mu_i$ for $i > \max(k, l)$, *a fortiori* for $i > \max(j_m, j'_m)$. Suppose for definiteness that $j_m \leqslant j'_m$, so that $\max(j_m, j'_m) = j'_m$. We must prove that $\alpha_{j'_m} = \mu_{j'_m}$. Our hypothesis that $\eta_{j_m} = \eta_{j_m+1} = \cdots = \eta_{j'_m-1} = \bigcirc$ implies that $\alpha_{j_m} = \alpha_{j_m+1} = \cdots \alpha_{j'_m}$, and so we get $\alpha_{j_m} = \mu_{j'_m}$. The case $j_m \geqslant j'_m$ is again similar. $\square$

We now introduce certain operations on signatures that give rise to concurrences.

DEFINITION 16.3 (Γ- and Δ-swaps) *Let σ and τ be subsignatures of η. We say that τ is obtained from σ by a Γ-swap at $i - 1, i$ if*

$$\sigma_j = \tau_j \quad \text{for all } j \neq i-1, i, \quad \sigma_{i-1} = *, \sigma_i = \square, \quad \tau_{i-1} = \bigcirc, \tau_i = *,$$

and by a Δ-swap at $i - 1, i$ if

$$\sigma_j = \tau_j \quad \text{for all } j \neq i-1, i, \quad \sigma_{i-1} = \square, \sigma_i = *, \quad \tau_{i-1} = *, \tau_i = \bigcirc.$$

DEFINITION 16.4 (Γ- and Δ-admissibility) *We say that a subset*

$$\Sigma = \{j_1, j_2, \cdots, j_m\}$$

of $\{1, 2, 3, \cdots, d\}$ is Γ-admissible for σ if

$$\mathcal{C}_\Gamma(\sigma) \subset \Sigma \subset \mathcal{C}_\Gamma(\sigma) \cup \mathcal{K}_\Gamma(\sigma),$$

and similarly it is Δ-admissible if $\mathcal{C}_\Delta(\sigma) \subset \Sigma \subset \mathcal{C}_\Delta(\sigma) \cup \mathcal{K}_\Delta(\sigma)$.

PROPOSITION 16.5 **(Swapped data concur)** *(a) Suppose τ is obtained from a Γ-swap at $i-1, i$. Assume that $i \notin \Sigma$. Let $0 < j_1 < j_2 < \cdots < j_l \leqslant d$ be a sequence such that $j_m \neq i$ for all m. Let*

$$j'_m = \begin{cases} j_m & \text{if } j_m \neq i-1; \\ i & \text{if } j_m = i-1. \end{cases}$$

Then (σ, Σ) and (τ, Σ') concur, where $\Sigma = \{j_1, \cdots, j_m\}$ and $\Sigma' = \{j'_1, \cdots, j'_m\}$. Moreover Σ is Γ-admissible for σ if and only if Σ' is Γ-admissible for τ.

(b) Suppose that τ is obtained from a Δ-swap at $i-1, i$. Assume that $i \notin \Sigma$. Let $0 < j_1 < j_2 < \cdots < j_l \leqslant d$ be a sequence such that $j_m \neq i$ for all m. Let

$$j'_m = \begin{cases} j_m & \text{if } j_m \neq i+1; \\ i & \text{if } j_m = i+1. \end{cases}$$

Then (σ, Σ) and (τ, Σ') concur, where $\Sigma = \{j_1, \cdots, j_m\}$ and $\Sigma' = \{j'_1, \cdots, j'_m\}$. Moreover Σ is Δ-admissible for σ if and only if Σ' is Δ-admissible for τ.

Proof. This is straightforward to check from the definitions of concurrence and admissibility. One point merits further discussion. Suppose we are in case (a) for definiteness. If Σ is Γ-admissible, then according to the definition (15.1), $i - 1 \in \mathcal{C}_\Gamma(\sigma) \subseteq \Sigma$ if $\sigma_0, \ldots, \sigma_{i-2}$ are not all $\circ$. In this case, $i \in \mathcal{C}_\Gamma(\tau) \subseteq \Sigma'$. If instead, $\sigma_0 = \cdots = \sigma_{i-2} = \circ$, then $i - 1 \notin \Sigma$ but then under the Γ-swap, $\tau_0 = \cdots = \tau_{i-1} = \circ$ and so $i \notin \Sigma'$. $\qquad\square$

If the hypotheses of Proposition 16.5 are satisfied we say that (τ, Σ') is obtained from (σ, Σ) by a Γ-*swap* (or Δ-*swap*).

Let us define an equivalence relation on the set of pairs (σ, Σ), where σ is a subsignature of η and Σ is a Γ-admissible subset of $\{1, 2, \cdots, d\}$.

DEFINITION 16.6 **(Γ- and Δ-packs)** *We write $(\sigma, \Sigma) \sim_\Gamma (\tau, \Sigma)$ if (τ, Σ) can be obtained by a sequence of Γ-swaps or inverse Γ-swaps. We call an equivalence class a Γ-pack; and Δ-packs are defined similarly.*

LEMMA 16.7 *Each Γ-pack or Δ-pack contains a unique element with a maximal number of $\circ$'s. Within the pack, this unique element (σ, Σ) is characterized as follows.*

Γ-*pack:*   *Whenever $\eta_{i-1}\eta_i = \circ\,\square$ and $\sigma_{i-1}\sigma_i = *\,\square$ we have $i \in \Sigma$,*

Δ-*pack:*   *Whenever $\eta_{i-1}\eta_i = \square\,\circ$ and $\sigma_{i-1}\sigma_i = \square\,*$ we have $i \in \Sigma$.*

Proof. If $\eta_{i-1}\eta_i = \circ\,\square$ and $\sigma_{i-1}\sigma_i = *\,\square$ then a Γ-swap is possible at $i-1, i$ if and only if $i \notin \Sigma$. Indeed, the fact that Σ is Γ-admissible for σ means that $i - 1 \in \Sigma$. This assertion therefore follows from Proposition 16.5.

Clearly the element maximizing the number of $\circ$'s is obtained by making all possible swaps. The statements are now clear for the Γ-pack, and for the Δ-pack they are similar. $\qquad\square$

DEFINITION 16.8 **(Origins)** *We call the unique element with the greatest number of $\circ$'s the origin of the pack. We say that (σ, Σ) is a Γ-origin if it is the origin of its Γ-pack, and Δ-origins are defined the same way.*

As we have explained, our goal is a bijection ψ between the Γ-packs and the Δ-packs. It is enough to exhibit a bijection between their origins. Let (σ, Σ) be the origin of a Γ-pack and denote $\psi(\sigma, \Sigma) = (\sigma', \Sigma')$. We can define σ' immediately. To obtain σ', we break η (which involves only $\square$'s and $\bigcirc$'s) into maximal strings of the form $\bigcirc \cdots \bigcirc$ and $\square \cdots \square$, and we prescribe σ' on these ranges.

- ($\bigcirc$'s in σ **reflect across the midpoint of the string of** $\bigcirc$'s in η) Suppose that $\eta_h, \ldots, \eta_k$ is a maximal string of consecutive $\bigcirc$'s in η (so $\eta_{h-1}, \eta_{k+1} \neq \bigcirc$). If $h \leqslant i \leqslant k$ then $\sigma_i' = \sigma_{h+k-i}$.

- (**Distinguished $\square$'s in σ slide one index leftward**) Suppose that $\sigma_h \ldots \sigma_k$ is a maximal string of consecutive $\square$'s in σ (so $\sigma_{h-1}, \sigma_{k+1} \neq \square$). Let $h \leqslant i \leqslant k$ be the smallest element of Σ in this range, or if none exists, let $i = k + 1$. Then if $\eta_{h-1} = \square$ and $\sigma_{h-1} = *$ then $\psi(\sigma) = \sigma'$ has $\sigma_{h-1}' = \cdots \sigma_{i-2}' = \square$, $\sigma_{i-1}' = *$, and $\sigma_i' = \cdots = \sigma_k' = \square$. If either $\eta_{h-1} = \bigcirc$ or $\sigma_{h-1} \neq *$, then ψ leaves the string of $\square$'s in σ unchanged.

The last rule merits further explanation. Since σ is a subsignature of η, the maximal chain $\sigma_h \cdots \sigma_k$ of boxes in σ is contained in a (usually longer) maximal chain of boxes $\eta_l \eta_{l+1} \cdots \eta_m$ within η; thus $l \leqslant h$ and $m \geqslant k$ and the range from l to m is thus broken up into smaller ranges of which $\sigma_h \ldots \sigma_k = \square \cdots \square$ is one. We assume that $\sigma_{h-1} = *$ and that $\eta_{h-1} = \square$. In this case we will modify $\sigma_h \cdots \sigma_k$. But if the condition that $\sigma_{h-1} = *$ and that $\eta_{h-1} = \square$ is not met, we leave it unchanged – and the condition will be met if and only if $h > l$. Then with i as in the second rule above, we make the following shift:

$$\left\{ \begin{matrix} \sigma_{h-1} & \sigma_h & \cdots & \sigma_{i-1} & \cdots & \sigma_k \\ * & \square & \cdots & \square & \cdots & \square \end{matrix} \right\} \longrightarrow \left\{ \begin{matrix} \sigma_{h-1}' & \sigma_h' & \cdots & \sigma_{i-1}' & \cdots & \sigma_k' \\ \square & \square & \cdots & * & \cdots & \square \end{matrix} \right\}.$$

$$(16.3)$$

It is useful to divide up the nodal signature η into blocks of consecutive $\square$'s alternating with blocks of consecutive $\bigcirc$'s (where a block might consist of just one of these characters), for example,

$$\eta = (\eta_0, \eta_1, \ldots, \eta_7) = (\square, \bigcirc, \bigcirc, \underbrace{\square, \square, \square}_{\text{block of } \square\text{'s}}, \bigcirc, \bigcirc,).$$

Formally, a $\square$-*block* is a maximal consecutive set $B = \{h, h+1, \cdots, k\}$ such that η_i are all $\square$'s, and $\bigcirc$-*blocks* are defined similarly. The map ψ can be understood according to what it does to the indices of σ contained within each of these blocks (and no two indices from different blocks interact under ψ). In particular, the number of $*$'s in σ contained within a block of η is preserved under ψ. We use this fact repeatedly in the proofs, as it often implies that it is enough to work locally within a block of $\square$'s or $\bigcirc$'s.

We have not yet described what ψ does to Σ. The next result will make this possible. Define

$$\begin{aligned} P_\sigma(u) &= |\{j \geqslant u \mid \sigma_j = *\}|, \\ Q_\sigma(u) &= |\{j \geqslant u \mid \sigma_j = \square\}|. \end{aligned}$$

If $u, v \in \{1, 2, \cdots, d\}$ then we say that the pair (u, v) is *equalized* for σ and σ' if

$$P_\sigma(u) = P_{\sigma'}(v), \qquad Q_\sigma(u) = Q_{\sigma'}(v). \tag{16.4}$$

LEMMA 16.9 *Let σ, σ' be signatures with $\psi(\sigma') = \sigma$.*

(i) If $1 \leqslant u \leqslant d$ and $\eta_u \neq \eta_{u-1}$ then (u, u) is equalized.

(ii) Suppose that B is a $\bigcirc$-block and that $u \in B$ such that $\sigma_u = *$. Assume that $\sigma_j \neq \bigcirc$ for some $j < u$. Then there exists $0 < v \in B$ such that $\sigma'_v = *$ and (u, v) is equalized.

(iii) Suppose that B is a $\bigcirc$-block and that $v \in B$ such that $\sigma'_v = *$ but $\sigma'_{v-1} \neq \bigcirc$. Then there exists $0 < u \in B$ such that (u, v) is equalized, $\sigma_u = *$ and $\sigma_j \neq \bigcirc$ for some $j < u$.

(iv) Given i as in the second rule for ψ on signatures, the pair (i, i) is equalized.

The condition in (ii) and (iii) that $\sigma_u = *$ and $\sigma_j \neq \bigcirc$ for some $j < u$ means that $u \in \mathcal{C}_\Gamma(\sigma)$.

Proof. Part (i) follows from the fact that u is at the left edge of a block when $\eta_u \neq \eta_{u-1}$. Indeed, if B is a $\square$- or $\bigcirc$-block then σ and σ' have the same number of $*$'s and $\square$'s in B. Since u is at the left edge of a block, the accumulated numbers of $*$ and $\square$ in that block and those to the right are the same for σ and σ' and so (u, u) is an equalized pair for σ and σ'.

To prove (ii), observe that the number of $*$ in the $\bigcirc$-block $B = \{h, h+1, \cdots, k\}$ are the same, and $(k+1, k+1)$ is an equalized pair (or else $k = d$), so counting from the right, if σ_u is the r-th $*$ within the block, we can take σ'_v to be the r-th $*$ for σ' in the block, and we have equalization. The hypothesis that $\sigma_u \neq \bigcirc$ for some $j < u$ guarantees that either B is not the first block, or that σ_u is not the leftmost $*$ in the block, so $v > 0$.

To prove (iii), we argue the same way, and the only thing to be checked is that $j > 0$ and that $\sigma_j \neq \bigcirc$ for some $j < u$. This follows from the assumption that $\sigma'_{v-1} \neq \bigcirc$, since if $\sigma'_{v-1} = \square$ then B is not the first block, while if $\sigma'_{v-1} = *$ then σ'_v is not the first $*$ in B for σ', hence also not the first $*$ in B for σ.

For (iv), the $\square$-block containing i can be broken up into segments of the form $*\square \cdots \square$ as in the left-hand side of (16.3) and possibly an initial string consisting entirely of $\square$'s. According to the second rule for ψ on signatures, the image of each such segment under ψ also contains exactly one $*$ (excluding the possible initial string of $\square$'s without $*$'s) and the same number of $\square$'s. As i occurs to the right of both the $*$ in σ and σ' in the respective segments as depicted in (16.3), it is thus clear that (i, i) is an equalized pair. $\qquad \square$

PROPOSITION 16.10 (Concurrence of origins) *Let (σ, Σ) be the origin of a Γ-pack, and let $\sigma' = \psi(\sigma)$. Given any $j \in \Sigma$ we can associate a corresponding index $j' \in \Sigma'$ as follows. There exists a unique $1 \leqslant t = t(j) \leqslant d$ such that (j, t) is an equalized pair, and such that $\sigma'_t \neq \bigcirc$. Define $j' = \psi(j)$ so that $j' - 1$ is the largest index $< t(j)$ such that $\sigma'_{j'-1} \neq \bigcirc$. Then the $\Sigma' = \{\psi(j)|j \in \Sigma\}$ is Δ-admissible for σ', and in fact (σ', Σ') is a Δ-origin. Moreover the pairs (σ, Σ) and (σ', Σ') concur. The map $\psi : (\sigma, \Sigma) \longmapsto (\sigma', \Sigma')$ is a bijection from the set of Γ-origins to the set of Δ-origins.*

Before proving this, we give several examples.

1. If $\eta = (\circ, \circ, \circ, \circ, \circ, \circ, \square)$,

$$\psi((\circ, \circ, *, \circ, *, *, *), \{4, 5, 6\}) \mapsto ((*, *, \circ, *, \circ, \circ, *), \{1, 2, 4\})$$

Indeed, ψ reflects all entries in the initial block of 6 $\circ$'s in η. In the block consisting of a single $\square$ at the end of η, σ contains no $\square$'s and so σ' agrees with σ on this block. The reader will check that $t(6) = 4, t(5) = 2$, and $t(4) = 1$. Thus Σ' is as defined in the Proposition. To check that Σ is Γ-admissible for σ, note that $\mathcal{C}_\Gamma(\sigma) = \{4, 5, 6\}$ and $\mathcal{K}_\Gamma(\sigma) = \varnothing$ so indeed Σ is to be of form $\mathcal{C}_\Gamma(\sigma) \cup \Phi$ where Φ is a (possibly empty) subset of $\mathcal{K}_\Gamma(\sigma)$. Moreover, we wanted to ensure that Σ' is of the form $\Sigma' = \mathcal{C}_\Delta(\sigma') \cup \Phi'$ where $\Phi' \subseteq \mathcal{K}_\Delta(\sigma')$. Referring back to the definitions of these sets in (15.9) and (15.10), we see that $\mathcal{C}_\Delta(\sigma') = \{1, 2, 4\}$ so we satisfy the necessary condition. (For the record, $\mathcal{K}_\Delta(\sigma') = \varnothing$ in this case.) Finally, no Γ-swaps or Δ-swaps are possible so (σ, Σ) is a Γ-origin and (σ', Σ') is a Δ-origin.

2. If $\eta = (\square, \square, \circ, \circ, \square, \square, \circ, \square, \circ)$,

$$\psi((\square, *, \circ, \circ, \square, \square, *, *, \circ), \{1, 4, 5, 6, 7\}) \mapsto$$
$$((\square, *, \circ, \circ, \square, \square, *, *, \circ), \{1, 2, 5, 6, 7\})$$

Note that there is no change in the signature from σ to σ' as no $\square$'s can move left in the strings of $\square$'s contained in η, and reflection in strings of $\circ$'s leaves these strings unchanged. The index sets are more interesting. From the definitions, we compute that $\mathcal{C}_\Gamma(\sigma) = \{1, 4, 6, 7\}$, $\mathcal{K}_\Gamma(\sigma) = \{5\}$, $\mathcal{C}_\Delta(\sigma') = \{2, 7\}$, and $\mathcal{K}_\Delta(\sigma') = \{1, 5, 6\}$. This illustrates that these sets may have very different cardinalities. Thus the sets Σ and Σ' are admissible.

3. If $\eta = (\circ, \square, \circ, \square, \square, \square, \square, \circ)$,

$$\psi((\circ, \square, \circ, *, \square, \square, \square, *), \{3, 6, 7\}) \mapsto$$
$$((\circ, \square, \circ, \square, \square, *, \square, *), \{2, 6, 7\})$$

Here, the blocks of circles are all of length 1, so $*$'s and $\circ$'s in σ within these blocks do not change under ψ in σ'. We have $\sigma_2 = \square$, but $2 \in \mathcal{N}_\Gamma(\sigma) \subseteq \Sigma$ so this $\square$ remains fixed in σ'. In the block of four $\square$'s, we see σ contains 3 $\square$'s. The smallest index from this string that is in Σ is 6, corresponding to the last $\square$. So the first two $\square$'s move left, and the third remains fixed. Again, from the definitions, we compute that $\mathcal{C}_\Gamma(\sigma) = \{3, 7\}$, $\mathcal{K}_\Gamma(\sigma) = \{4, 5, 6\}$, $\mathcal{C}_\Delta(\sigma') = \{2, 6\}$, and $\mathcal{K}_\Delta(\sigma') = \{4, 5, 7\}$, so Σ and Σ' are admissible.

Proof. The first thing to check is that if $j \in \Sigma$ we may find v such that (j, v) is an equalized pair. (If $j \notin \Sigma$ this may not be true.) There may be several possible v's, if σ' has $\circ$'s in the vicinity, and t will be the smallest. So the existence of v is all that needs to be proved – the condition that $\sigma'_t \neq \circ$ has the effect of selecting the smallest, so that t will be uniquely determined.

If B is a $\circ$-block, then the existence of v is guaranteed by Lemma 16.9. If B is a $\square$-block, then B can be broken into segments in which σ and σ' are as follows.

There is an initial segment (possibly empty) of $\square$'s that is common to both σ and σ', and the remaining segments look like this:

$$\left\{\begin{array}{cccccc} \sigma_l & \sigma_{l+1} & \sigma_{l+1} & \cdots & \sigma_{m-1} & \sigma_m \\ * & \square & \square & \cdots & \square & \square \end{array}\right\} \xrightarrow{\psi} \left\{\begin{array}{ccccc} \sigma'_l & \sigma'_{l+1} & \cdots & \sigma'_{m-1} & \sigma'_m \\ \square & \square & \cdots & \square & * \end{array}\right\}.$$

We claim that the only possible element of Σ in $\{l, l+1, \cdots, m\}$ is l. The reason is that if there was an element i of Σ in the range $\{l+1, \cdots, m\}$ then the prescription for σ' would move the $*$ to $i - 1$, and this is not the case. Now $(m+1, m+1)$ is an equalized pair (or $m = d$) and it follows that (l, l) is also equalized. So we have the case $j = l$, and then we can take $v = l$ also. It is easy to see that if j lies in the initial segment (if nonempty) that consists entirely of $\square$'s then we may take $u = j$ in this case also.

This proves that t satisfying (16.4) exists.

We will make use of the following observation.

If $\eta_{l'-1}\eta_{l'} = \circ\square$ and $\sigma'_{j'-1} = *$ for some $j' \leqslant l'$ then $l' \in \Sigma$ and $t(l') = l'$.
$$(16.5)$$

To prove this, note that if $\sigma_{l'} = *$ or if $\sigma_{l'} = \square$ and $\sigma_{l'-1} = \circ$ then $l' \in \mathcal{C}_\Gamma(\sigma) \subseteq \Sigma$. The fact that $\sigma_i \neq \circ$ for some $i < l'$, needed here for the definition of $\mathcal{C}_\Gamma(\sigma)$, may be deduced from the fact that $\sigma'_{j'-1} \neq \circ$ since it means that the $\circ$-block containing j' either is not the first block, or else contains some $*$'s for σ' and hence also for σ. Since $\eta_{l'-1}\eta_{l'} = \circ\square$ this leaves only the case $\sigma_{l'-1}\sigma_{l'} = *\square$, and in this case $l' \in \Sigma$ follows from the fact that σ is a Γ-origin by Lemma 17.1. Now (l', l') is an equalized pair by Lemma 16.9 (i), and so $t(l') = l'$. This proves (16.5).

Now we need to check that $\Sigma' = \{j'_1, j'_2, \cdots\}$ is Δ-admissible, that is, $\mathcal{C}_\Delta(\sigma') \subseteq \Sigma' \subseteq \mathcal{C}_\Delta(\sigma') \cup \mathcal{K}_\Delta(\sigma')$. That $\Sigma' \subseteq \mathcal{C}_\Delta(\sigma') \cup \mathcal{K}_\Delta(\sigma')$ is almost immediate, as the set $\mathcal{C}_\Delta(\sigma') \cup \mathcal{K}_\Delta(\sigma')$ contains every index j' with $\sigma'_{j'-1} = \square$ or $*$, so long as $\sigma'_{i'} \neq \circ$ for some $i' \geqslant j'$. Since each element $\psi(j) = j' \in \Sigma'$ with $j \in \Sigma$ has $\sigma'_{j'-1} = \square$ or $*$, we need only check that $\sigma'_{i'} \neq \circ$ for some $i' \geqslant j'$. This is clear since $j \in \mathcal{C}_\Gamma(\sigma) \cup \mathcal{K}_\Gamma(\sigma)$ so $P_\sigma(j)$ or $Q_\sigma(j)$ is positive, and hence $P_{\sigma'}(t)$ or $Q_{\sigma'}(t)$ is positive, and $j' \leqslant t$.

We next show that Σ' contains $\mathcal{C}_\Delta(\sigma')$. Thus to each $j' \in \mathcal{C}_\Delta(\sigma')$ we must find $j \in \Sigma$ such that $\psi(j) = j'$.

First assume that $\sigma'_{j'} = \circ$. By definition of $\mathcal{C}_\Delta(\sigma')$ we have $\sigma'_{j'-1} \neq \circ$. Also by definition of $\mathcal{C}_\Delta(\sigma')$ there will be some $l' > j'$ such that $\sigma'_{l'} \neq \circ$. Let l' be the smallest such value. Suppose that $\eta_{l'} = \circ$. Then $\sigma'_{l'} = *$. Since l' is the smallest value $> j'$ such that $\sigma'_{l'} \neq \circ$ we have $\sigma'_i = \circ$ and hence $\eta'_i = \circ$ for $j' \leqslant i < l'$ and so the entire range $j' \leqslant i \leqslant l'$ is contained within the same $\circ$-block B. By Lemma 16.9 (iii) there exists $j \in B$ such that $\sigma_j = *$ and (j, j') is an equalized pair, and moreover, $\sigma_i \neq \circ$ for some $i < j$. Thus $j \in \mathcal{C}_\Gamma(\sigma)$ so $j \in \Sigma$ and $t(j) = j'$, so $\psi(j) = j'$ (because $\sigma'_{j'-1} \neq \circ$). Thus we may assume that $\eta_{l'} = \square$. We note that $\eta_{l'-1} = \circ$ since $\sigma'_{l'-1} = \circ$. Thus $l' \in \Sigma'$ and $t(l') = l'$ by (16.5). Since $\sigma'_{j'} = \sigma'_{j'-1} = \ldots = \sigma'_{l'-1} = \circ$ but $\sigma'_{j'-1} \neq \circ$ we have $\psi(l') = j'$. This finishes the case $\sigma'_{j'} = \circ$.

Next suppose that $\sigma'_{j'} \neq \circ$. Then $\sigma'_{j'-1} \neq *$ since $j' \in \mathcal{C}_\Delta(\sigma')$. If $\eta_{j'} = \circ$ then $\sigma'_{j'}$ must be $*$. The assumption that $j' \in \mathcal{C}_\Delta(\sigma')$ then implies that $\sigma'_{j'-1} = *$

also and so Lemma 16.9 (iii) implies that $t(j) = j'$ for some j in the same $\circ$-block as j', with $j \in \mathcal{C}_\Gamma(\sigma) \subseteq \Sigma$. Then since $\sigma'_{j'-1} \neq \circ$ we have $\psi(j) = j'$. Thus we may assume that $\eta_{j'} = \square$. In this case we will show that $j' \in \Sigma$ and $\psi(j') = j'$ If $\eta_{j'-1}\eta_{j'} = \square\square$ then since $\sigma'_{j'-1} = *$ it follows from the description of σ' in $\square$-blocks (see (16.3)) that $j' \in \Sigma$, and by Lemma 16.9 (iii), (j', j') is an equalized pair, so $t(j') = j'$ and so since $\sigma'_{j'-1} = *$ it follows that $\psi(j') = j'$. On the other hand if $\eta_{j'-1}\eta_{j'} = \circ\square$ then we still have $j' \in \Sigma$ by (16.5), and since $\sigma'_{j'-1} = *$ we have $\psi(j') = j'$. This completes the proof that Σ' contains $\mathcal{C}_\Delta(\sigma')$.

Now we know that Σ' is Δ-admissible for σ'. Next, we show that (σ', Σ') is a Δ-origin. We must show that if $\eta_{j'-1}\eta_{j'} = \square\circ$ and $\sigma'_{j'-1}\sigma'_{j'} = \square*$ then $j' \in \Sigma'$. It follows from Lemma 16.9 that there exists j in the same $\circ$-block as j' such that $\sigma_j = *$ and $j \in \mathcal{C}_\Gamma(\sigma)$, and (j, j') is an equalized pair. Then $t(j) = j'$ and since $\sigma_{j'-1} \neq \circ$ we have $\psi(j) = j'$. Thus $j' \in \Sigma'$.

Next we observe that (σ, Σ) and (σ', Σ') concur. To see this, observe first that if $j \in \Sigma$ and $j' = \psi(j) \in \Sigma'$, then (j, j') is an equalized pair. This implies that the two sets (16.1) have the same cardinality (with $\tau = \sigma'$). Moreover, if j is in a $\circ$-block, then j' is in the same block, while if j is in a $\square$-block then $j' = j$ with the exception that if j is the left-most element of a $\square$-block, then j' can lie in the $\circ$-block to the left. These considerations show that $\eta_i = \circ$ when (16.2) is satisfied. Therefore σ and σ' concur.

We see that ψ maps Γ-origins to Δ-origins. To establish that it is a bijection between Γ-origins and Δ-origins, we first note show the map ψ from Γ-origins to Δ-origins is injective. Indeed, we can reconstruct σ and Σ from σ' and Σ' as follows. On $\circ$-blocks, the reconstruction is straightforward – the signature is just reversed on each Δ-block, and the elements of Σ within a Δ-block are just the values where $\sigma_j = *$, except that if $\sigma_i = \circ$ for all $i < j$ then j is omitted from σ. On $\square$-blocks, the reconstruction of Σ must precede the reconstruction of σ. It follows from the preceding discussion that on the intersection of Σ with a $\square$-block, ψ is the identity map except that if the very first element of the block is in Σ, ψ can move it into the preceding $\circ$-block. Thus if $j \in \{1, 2, \cdots, d\}$ and $\eta_j = \square$ we can tell if j is in Σ as follows. If j is not the first element on its block then $j \in \Sigma$ if and only if $j \in \Sigma'$. If j is the first element, then $j \in \Sigma$ if and only if $j \in \Sigma'$ or (from the definition of $\mathcal{C}_\Gamma$) if $\sigma_{j-1} = \circ$ – and we recall that the signature is already known on $\circ$-blocks. Once Σ is known on $\square$-blocks, σ can be reconstructed by reversing the process that gave us σ'.

Since the map ψ is injective, we need only check that the number of Δ-origins equals the number of Γ-origins. We extend ψ to a larger set by including η as part of the data: let Ω_Γ be the set of all triples (η, σ, Σ) such that η is a nodal signature, σ a subsignature, and Σ a Γ-origin for σ; and similarly we define Ω_Δ. Then ψ gives an injection $\Omega_\Gamma \longrightarrow \Omega_\Delta$. It will follow that ψ is a bijection if we show that the two sets have the same cardinality. A naive bijection between the two sets can be exhibited as follows. Let (η, σ, Σ) be given. Define $(\tilde{\eta}, \tilde{\sigma}, \tilde{\Sigma})$ by $\tilde{\eta}_i = \eta_{d-i}$, $\tilde{\sigma}_i = \sigma_{d-i}$, and $\tilde{\Sigma} = \{d+1-j | j \in \Sigma\}$. Note that η and σ are reversed in the range from 0 to d, while Σ is reversed in the range from 1 to d. Then $(\tilde{\sigma}, \tilde{\Sigma})$ is a Δ-origin if and only if (σ, Σ) is a Γ-origin, and so $|\Omega_\Gamma| = |\Omega_\Delta|$. $\quad\square$

Chapter Seventeen

Conclusion of the Proof

In Chapter 15 we reduced the proof of Theorem 1.2 to Statement G, given at the end of that chapter, and we now have the tools to prove it.

LEMMA 17.1 *The cardinality of each Γ-pack or Δ-pack is a power of 2.*

Proof. In a Γ-swap $*\square$ is replaced by $\circ *$ in the signature. Since both signatures are subsignatures of η, this means that η has $\circ \square$ at this location. From this it is clear that if a Γ-swap is possible at $i - 1, i$ then no swap is possible at $i - 2, i - 1$, or $i, i + 1$, and so the swaps are independent. Thus the cardinality of the pack is a power of 2. $\qquad\square$

Given an origin (σ, Σ) for a Γ-equivalence class, define $p_\Gamma(\sigma, \Sigma) = k$ where 2^k is the cardinality of the Γ-pack to which the representative (σ, Σ) belongs. We may similarly define p_Δ for Δ-packs.

PROPOSITION 17.2 *Let (σ, Σ) be an origin for a Γ-equivalence class. Then*

$$p_\Gamma(\sigma, \Sigma) = \big| \{i \in \{1, 2, \cdots, d\} \mid (\sigma_{i-1}, \sigma_i) = (\circ, *), \eta_i = \square\} \big|. \qquad (17.1)$$

Proof. Recall that elements of a Γ-pack differ by a series of Γ-swaps from (τ, T) to (σ, Σ), which change $\tau_{i-1}, \tau_i = *, \square$ to $\sigma_{i-1}, \sigma_i = \circ, *$ provided $i \notin T$. Hence $p_\Gamma(\sigma, \Sigma)$ is clearly at most the number of indices satisfying the condition on the right-hand side of (17.1).

Given an origin (σ, Σ), let $\tau \subseteq \eta$ be any subsignature possessing an $(*, \square)$ at (τ_{i-1}, τ_i) where σ has a $(\circ, *)$ at (σ_{i-1}, σ_i). Let T be the set of indices obtained from Σ by replacing each such $i \in \Sigma$ by $i - 1$ (and leaving all other indices unchanged). We claim that $(\tau, T) \sim_\Gamma (\sigma, \Sigma)$. By our discussion above, it suffices to show that $i \notin T$. Indeed, by assumption, $(\sigma_i, \sigma_{i+1}) \neq (\circ, *)$ so $i+1$ is not changed to i from Σ to T according to our rule. Moreover, $i \in \Sigma$ is sent to $i - 1 \in T$. Hence (17.1) follows. $\qquad\square$

LEMMA 17.3 *Let E_Γ be a Γ-pack with origin (σ, Σ). Let $\Sigma = C_\Gamma(\sigma) \cup \Phi$ with $\Phi \subseteq K_\Gamma(\sigma)$ and let $x = |\Phi|$. Then*

$$\sum_{(\sigma, \Sigma) \in E_\Gamma} (-1)^\epsilon \, V(a_\Gamma(\sigma) + x, a_\Gamma(\sigma) + k_\Gamma(\sigma)) =$$

$$(-1)^\epsilon V(a_\Gamma(\sigma) + x + p_\Gamma(\sigma, \Sigma), a_\Gamma(\sigma) + k_\Gamma(\sigma) + p_\Gamma(\sigma, \Sigma)),$$

where $\epsilon = \epsilon(\sigma)$ is the number of $\circ$ in σ and ϵ is the number of $\circ$ in σ.

Proof. It is easy to see that a Γ-swap does not change $a_\Gamma(\sigma)$, while it decreases $k_\Gamma(\sigma)$ by 1. Thus repeatedly applying the identity

$$V(a,b) - V(a, b+1) = V(a+1, b+1)$$

gives this result.   □

LEMMA 17.4 *Let E_Δ be a Δ-pack with origin (σ, Σ). Let $\Sigma = C_\Delta(\sigma) \cup \Phi$ with $\Phi \subseteq K_\Gamma(\sigma)$ and let $x = |\Phi|$. Then*

$$\sum_{(\sigma,\Sigma)\in E_\Delta} (-1)^\epsilon \, V(a_\Delta(\sigma) + x, a_\Delta(\sigma) + k_\Delta(\sigma)) =$$

$$(-1)^\epsilon V(a_\Delta(\sigma) + x + p_\Delta(\sigma, \Sigma), a_\Delta(\sigma) + k_\Delta(\sigma) + p_\Delta(\sigma, \Sigma)),$$

where $\epsilon = \epsilon(\sigma)$ is the number of $\circ$ in σ and ϵ is the number of $\circ$ in σ.

Proof. Similar to the proof of Lemma 17.3.   □

THEOREM 17.5 *Let ψ be the bijection on equivalence classes given above, let (σ, Σ) be a Γ-origin and let $\psi(\sigma, \Sigma) = (\sigma', \Sigma')$ be the corresponding Δ-origin. Write $\Sigma = C_\Gamma(\sigma) \cup \Phi$ with $\Phi \subseteq K_\Gamma(\sigma)$ and similarly, $\Sigma' = C_\Delta(\sigma') \cup \Phi'$ with $\Phi' \subseteq K_\Delta(\sigma')$. Then*

$$V(a_\Gamma(\sigma) + |\Phi| + p_\Gamma(\sigma, \Sigma), a_\Gamma(\sigma) + k_\Gamma(\sigma) + p_\Gamma(\sigma, \Sigma)) =$$

$$V(a_\Delta(\sigma') + |\Phi'| + p_\Delta(\sigma', \Sigma'), a_\Delta(\sigma') + k_\Delta(\sigma') + p_\Delta(\sigma', \Sigma')). \quad (17.2)$$

Proof. First we will prove the equality of the second components

$$a_\Gamma(\sigma) + k_\Gamma(\sigma) + p_\Gamma(\sigma, \Sigma) = a_\Delta(\sigma') + k_\Delta(\sigma') + p_\Delta(\sigma', \Sigma'). \quad (17.3)$$

Consider the left-hand side of (17.3). The quantities $k_\Gamma(\sigma)$ and $a_\Gamma(\sigma)$ are defined in Chapter 15 before Proposition 15.1, where the latter is further defined in terms of $n_\Gamma(\sigma)$ and $t_\Gamma(\sigma)$. Hence, the left-hand side of (17.3) is given by

$$2(d - (t_\Gamma - n_\Gamma)) + \left\{ \begin{array}{ll} -1 & \sigma_0 = \circ \\ 0 & \sigma_0 = \square \\ 1 & \sigma_0 = * \end{array} \right\}$$

$$+ \left\{ \begin{array}{ll} 1 & \sigma_d = \circ \\ 0 & \sigma_d \neq \circ \end{array} \right\} + |\{i \in [1,d] \mid \sigma_i = \square, \sigma_{i-1} \neq \circ\}|$$

$$+ |\{i \in [1,d] \mid (\sigma_{i-1}, \sigma_i) = (\circ, *), \eta_i = \square\}|,$$

where $[a,b]$ denotes $\{x \in \mathbb{Z} \mid a \leqslant x \leqslant b\}$. Now

$$2d - 2t_\Gamma + \left\{ \begin{array}{ll} -1 & \sigma_0 = \circ \\ 0 & \sigma_0 = \square \\ 1 & \sigma_0 = * \end{array} \right\} = 2d + 1 - 2BC(\sigma) + \left\{ \begin{array}{ll} 1 & \sigma_0 = \square \\ 0 & \sigma_0 \neq \square \end{array} \right\}$$

where $BC(\sigma) = |\{i \in [0,d] \mid \sigma_i \neq *\}|$ is the total number of boxes and circles in σ. Also the quantity $2n_\Gamma$ contributes a 2 for each $i \in [1,d]$ with $(\sigma_{i-1}, \sigma_i) = (\circ, \square)$. We may regard this 2 as contributing 1 for each $\square$ preceded by a $\circ$ and 1 for each $\circ$ followed by a $\square$. From this it follows that

$$2n_\Gamma + |\{i \in [1,d] \mid \sigma_i = \square, \sigma_{i-1} \neq \circ\}| +$$

$$|\{i \in [1,d] \mid (\sigma_{i-1}, \sigma_i) = (\circ, *), \eta_i = \square\}|$$

$$= |\{i \in [1,d] \mid \sigma_i = \square\}| + |\{i \in [0, d-1] \mid \sigma_i = \circ, \eta_{i+1} = \square\}|.$$

Combining terms, we see that the left-hand side of (17.3) is the sum of the two terms

$$2d + 1 - 2BC(\sigma) + \left| \{i \in [0,d] \mid \sigma_i = \square\} \right| \tag{17.4}$$

and

$$\left| \{i \in [0, d-1] \mid \sigma_i = \circ, \eta_{i+1} = \square\} \right| + \left\{ \begin{array}{ll} 1 & \sigma_d = \circ \\ 0 & \sigma_d \neq \circ \end{array} \right\}. \tag{17.5}$$

Similarly, the right-hand side of (17.3) is the sum of the two terms

$$2d + 1 - 2BC(\sigma') + \left| \{i \in [0,d] \mid \sigma'_i = \square\} \right| \tag{17.6}$$

and

$$\left| \{i \in [1, d] \mid \sigma'_i = \circ, \eta_{i-1} = \square\} \right| + \left\{ \begin{array}{ll} 1 & \sigma'_0 = \circ \\ 0 & \sigma'_0 \neq \circ \end{array} \right\}. \tag{17.7}$$

Now, since the map ψ preserves the number of boxes and the number of circles in the signature, we have

$$BC(\sigma) = BC(\sigma')$$

and

$$\left| \{i \in [0,d] \mid \sigma_i = \square\} \right| = \left| \{i \in [0,d] \mid \sigma'_i = \square\} \right|.$$

Hence the quantities (17.4) and (17.6) are equal. The quantity (17.5) counts the number of $i \in [0, d]$ such that $\sigma_i = \circ$, $\eta_{i+1} \neq \circ$ (this includes the possibility that $i = d$ and η_{i+1} is not defined). But ψ reflects the entries of σ lying over strings of $\circ$'s in η. After doing so, each such i reflects to a $\circ$ in σ' that is preceded by a $\square$ (or is initial) in η. These are exactly the indices counted by (17.7). Hence they are equal.

This completes the proof of (17.3). To finish the proof of the theorem, we must show that

$$a_\Gamma(\sigma) + |\Phi| + p_\Gamma(\sigma, \Sigma) = a_\Delta(\sigma') + |\Phi'| + p_\Delta(\sigma', \Sigma').$$

By the construction of the bijection ψ, we have

$$c_\Gamma(\sigma) + |\Phi| = c_\Delta(\sigma') + |\Phi'|,$$

since these count the number of divisibility conditions, and this number is necessarily constant when the bijection obtains. In view of (17.3), it thus suffices to establish

$$c_\Gamma(\sigma) + k_\Gamma(\sigma) = c_\Delta(\sigma') + k_\Delta(\sigma'). \tag{17.8}$$

The case $\eta_i = \sigma_i = \circ$ for all $0 \leqslant i \leqslant d$ is trivial and we exclude it henceforth.

The quantity $c_\Gamma(\sigma) + k_\Gamma(\sigma)$ counts the number of $i \in [1, d]$ such that $\sigma_i = \square$ or $\sigma_i = *$ but $\sigma_0, \ldots, \sigma_{i-1}$ are not all $\circ$. We claim that (excluding the trivial case above)

$$c_\Gamma(\sigma) + k_\Gamma(\sigma) = \left| \{i \in [0, d] \mid \sigma_i = \square \text{ or } \sigma_i = *\} \right| - 1. \tag{17.9}$$

To check this, there are two cases. First, suppose $\sigma_0 = \square$ or $\sigma_0 = *$. Then the index 0 is counted in the first term on the right-hand side of (17.9) even though it is not in the range $1 \leqslant i \leqslant d$, but this is accounted for by subtracting 1 there. The indices $i \in [1, d]$ with $\sigma_i = \square$ or $\sigma_i = *$ are counted on both sides. Hence (17.9) holds. The other possibility is $\eta_0 = \sigma_0 = \circ$. The index $i = 0$ is not counted in the first term on the right-hand side of (17.9). However, σ begins with a $\circ$, and the first index i_0 such that $\sigma_{i_0} \neq \circ$ *is* counted in the first term on the right-hand side of (17.9). Subtracting 1 there makes up for the exclusion of the index i_0 on the left-hand side as it corresponds to a $\square$ or $*$ preceded by a nonempty initial string of $\circ$'s. The remaining indices $i > i_0$ such that $\sigma_i = \square$ or $\sigma_i = *$ are counted on both sides. Hence (17.9) is also true in this case.

Similarly, we have (again excluding the case that all $\sigma'_i = \circ$)

$$c_\Delta(\sigma') + k_\Delta(\sigma') = |\{i \in [0, d] \mid \sigma'_i = \square \text{ or } \sigma'_i = *\}| - 1.$$

But since the map ψ preserves the number of boxes and the number of stars in the signature, we conclude that (17.8) holds, and the theorem is proved. $\square$

This completes the proofs of Theorems 1.2 and 1.1.

Chapter Eighteen

Statement B and Crystal Graphs

Theorem 1.2 is now proved, but we have further important remarks to make, which will occupy the last three chapters. These chapters may be read independently of each other.

In this chapter, we will translate Statements A and B from Chapter 6 into Statements A′ and B′ in the language of crystal bases, and explain in this language how Statement B′ implies Statement A′.

We slightly generalize the definition (2.14) by considering more general reduced words Ω and defining

$$G_\Omega(v) = \prod_{b_i \in \mathrm{BZL}_\Omega(v)} \begin{cases} g(b_i) & \text{if } b_i \text{ is boxed but not circled in } \mathrm{BZL}_\Omega(v), \\ q^{b_i} & \text{if } b_i \text{ is circled but not boxed,} \\ h(b_i) & \text{if } b_i \text{ is neither circled nor boxed,} \\ 0 & \text{if } b_i \text{ is both boxed and circled.} \end{cases}$$

This definition is provisional since it assumes that we can give an appropriate definition of boxing and circling for Ω. We will only use this notation for certain words Ω, some of them not necessarily long, and certain v, and in every case we will explain what the boxing and circling rules are.

For the long words Ω_Γ and Ω_Δ defined by (2.9) and (2.10) we have $G_{\Omega_\Gamma}(v) = G_\Gamma(v)$ and $G_{\Omega_\Delta}(v) = G_\Delta(v)$. Thus by Theorem 2.5, Statement A can be paraphrased as follows.

Statement A′. *We have*

$$\sum_{\mathrm{wt}(v)=\mu} G_{\Omega_\Gamma}(v) = \sum_{\mathrm{wt}(v)=\mu} G_{\Omega_\Delta}(v).$$

We believe that if the correct definition of the boxing and circling decorations can be given, we could say that $\sum_{\mathrm{wt}(v)=\mu} G_\Omega(v)$ is independent of the choice of Ω. However the description of the boxing and circling might be different for Ω other than Ω_Δ and Ω_Γ, and we will limit our discussion to those two words. This need for caution may be related to assumptions required by Littelmann [57] in order to specify sets of BZL patterns associated to a particular "good" long word. Littelmann found that for particular choices of "good" decompositions, including $\Omega = \Omega_\Gamma, \Omega_\Delta$, one can easily compute explicit inequalities that describe a polytope whose integer lattice points parametrize the set of all BZL patterns in a highest weight crystal. The decoration rules are closely connected to the location of $\mathrm{BZL}_\Omega(v)$ in this polytope.

The crystal graph formulation in Statement A′ is somewhat simpler than its Gelfand-Tsetlin counterpart. In particular, in the formulation of Statement A, we had two different Gelfand-Tsetlin patterns $\mathfrak{T}$ and $\mathfrak{T}'$ that were related by the Schütz-enberger involution, but the equality in Statement A was further complicated because the involution changes the weight of the pattern. In the crystal graph formulation, different decompositions of the long element simply result in different paths from the same vertex v to the lowest weight vector.

We will explain how Statement A' can be proved inductively. First we must explain the interpretation of the short Gelfand-Tsetlin patterns $\mathfrak{t}$ and their associated preaccordions $\Gamma_\mathfrak{t}$ and $\Delta_\mathfrak{t}$ in the crystal language.

Removing all edges labeled either 1 or r from the crystal graph results in a disjoint union of crystals of Cartan type A_{r-2}. The root operators for one of these subcrystals have indices shifted – they are $f_2, \cdots, f_{r-1}$ and $e_2, \cdots, e_{r-1}$ – but this is unimportant. Each such subcrystal has a unique lowest weight vector, characterized by $f_i(v) = 0$ for all $1 < i < r$. If $v \in \mathcal{B}_\lambda$ we will say that v is a *short vector* if $f_i(v) = 0$ for all $1 < i < r$. Thus there is a bijection between these subcrystals and the short vectors.

Now consider the words

$$\omega_\Gamma = (1, 2, 3, \cdots, r-1, r, r-1, \cdots, 3, 2, 1)$$

and

$$\omega_\Delta = (r, r-1, r-2, \cdots, 3, 2, 1, 2, 3, \cdots, r-1, r).$$

These words are reduced by not long. Identifying the Weyl group with the symmetric group S_{r+1} and the simple reflections $\sigma_i \in W$ with transpositions $(i, i+1)$, these give reduced decompositions of the reflection $(1, r+1)$. That is, if $\omega = \omega_\Gamma$ or ω_Δ and

$$\omega = (b_1, \cdots, b_{2r-1})$$

then $\sigma_{b_1} \cdots \sigma_{b_{r+1}} = (1, r+1)$.

The following result interprets the preaccordions $\Gamma_\mathfrak{t}$ and $\Delta_{\mathfrak{t}'}$ of a short Gelfand-Tsetlin pattern, which have occupied so much space in this document, as paths in the crystal.

THEOREM 18.1 *Let v be a short vector, and let $\omega = \omega_\Gamma$ or ω_Δ. Then we have*

$$v \begin{bmatrix} b_1 & \cdots & b_{2r-1} \\ \omega_1 & \cdots & \omega_{2r-1} \end{bmatrix} v' \tag{18.1}$$

with $v' = v_{\text{low}}$. Moreover, the b_i satisfy the inequalities

$$b_1 \geqslant b_2 \geqslant \ldots \geqslant b_{r-1} \geqslant 0, \qquad b_r \geqslant b_{r+1} \geqslant \ldots \geqslant b_{2r-1} \geqslant 0.$$

Let $\mathfrak{t} = \mathfrak{t}(v)$ be the short Gelfand-Tsetlin pattern obtained by discarding all but the top three rows of $q_{r-1}\mathfrak{T}_v$. Then if $\omega = \omega_\Gamma$ we have in the notation of (6.9) with $d = r - 1$:

$$\Gamma_\mathfrak{t} = \left\{ \begin{matrix} b_r & & b_{r+1} & & b_{r+2} & & \cdots & & b_{2r-2} & & b_{2r-1} \\ & b_{r-1} & & b_{r-2} & & & \cdots & & b_2 & & b_1 \end{matrix} \right\}. \tag{18.2}$$

On the other hand if $\omega = \omega_\Delta$ *then in the notation (6.10)*

$$
\Delta_{t'} = \left\{ \begin{array}{cccccc} b_{2r-1} & b_{2r-2} & b_{2r-3} & \cdots & b_{r+1} & b_r \\ & b_1 & b_2 & \cdots & b_{r-2} & b_{r-1} \end{array} \right\}.
$$

$$(18.3)$$

If v_1 *and* v_2 *are two short vectors such that* $t(v_1)$ *and* $t(v_2)$ *are in the same short pattern prototype, then* $\mathrm{wt}(v_1) = \mathrm{wt}(v_2)$.

Proof. Let $\mathcal{B}_\mu$ be the A_{r-1} crystal containing v that is obtained from $\mathcal{B}_\lambda$ by deleting the r-labeled edges. We make use of the word

$$
\Omega_{\Delta,r-1} = (r-1, r-2, r-1, r-3, r-2, r-1, \cdots, 1, 2, 3, \cdots, r-1),
$$

which represents the long element of A_{r-1} and obtain a path

$$
v \left[\begin{array}{ccccccccc} 0 & 0 & 0 & \cdots & 0 & b_1 & b_2 & b_3 & \cdots & b_{r-1} \\ r-1 & r-2 & r-1 & \cdots & r-1 & 1 & 2 & 3 & \cdots & r-1 \end{array} \right] v'
$$

where the initial string of 0's is explained by the fact that $f_i v = 0$ when $2 \leqslant i \leqslant r-1$. Thus we could equally well write

$$
v \left[\begin{array}{ccccc} b_1 & b_2 & b_3 & \cdots & b_r \\ 1 & 2 & 3 & \cdots & r \end{array} \right] v'.
$$

By Proposition 2.2, v' is the lowest weight vector of $\mathcal{B}_\mu$, so $f_1 v' = \cdots = f_{r-1} v' = 0$. Next we make use of the word

$$
\Omega_\Gamma = (1, 2, 1, 3, 2, 1, \cdots, r, r-1, \cdots, 3, 2, 1)
$$

and apply it to v'. Again, the first f_i that actually "moves" v' is f_r, and so we obtain a path

$$
v' \left[\begin{array}{ccccccccc} 0 & 0 & 0 & \cdots & 0 & b_r & b_{r+1} & b_{r+2} & \cdots & b_{2r-1} \\ 1 & 2 & 1 & \cdots & 1 & r & r-1 & r-2 & \cdots & 1 \end{array} \right] v_{\mathrm{low}},
$$

which we could write

$$
v' \left[\begin{array}{ccccc} b_r & b_{r+1} & b_{r+2} & \cdots & b_{2r-1} \\ r & r-1 & r-2 & \cdots & 1 \end{array} \right] v_{\mathrm{low}}.
$$

Splicing the two paths we get (18.1).

Next we prove (18.2). We note that the top row of Γ_t depends only on the top two rows of t, which are the same as the top two rows of $\mathfrak{T} = \mathfrak{T}_v$ since q_{r-1} does not affect these top two rows and t consists of the top three rows of $q_{r-1}t$. The top row of Γ_t is obtained from the top two rows of t by the right-hand rule (see Chapter 1), and so it agrees with the top row of $\Gamma_{\mathfrak{T}}$.

Now we regard v' as an element of the crystal $\mathcal{B}_\mu$ and apply the word $\Omega_{\Delta,r-1}$. We see that $b_1, \cdots, b_r$ are the top row of $\Delta(q_{r-1}\mathfrak{T}_{r-1})$ where $\mathfrak{T}_{r-1}$ is the Gelfand-Tsetlin pattern obtained by discarding the top row of $\mathfrak{T}$. Now the top two rows of $q_{r-1}\mathfrak{T}_{r-1}$ are the middle and bottom rows of t, which in Γ_t is read by the left-hand rule, which is the same as $\Delta(q_{r-1}\mathfrak{T}_{r-1})$. It follows that $b_1, \cdots, b_r$ form the top row of Γ_t, as required. This proves (18.2).

It remains for us to prove (18.3). As in Proposition 2.2 we can make use of ϕ_v, which interchanges the words ω_Γ and ω_Δ. Using (2.12) and arguing as at the end of Proposition 2.2 we see that the right-hand side of (18.3) equals Γ_u^{rev}, where u is the short Gelfand-Tsetlin pattern obtained by taking the top three rows of $-q_{r-1}q_r\mathfrak{T}_v^{rev}$. Now we make use of (7.1) in the form $q_{r-1}q_r = q_{r-2}t_rq_{r-1}$ to see that u is the short Gelfand-Tsetlin pattern obtained by taking the top three rows of $-q_{r-2}t_rq_{r-1}\mathfrak{T}_v^{rev}$, and since q_{r-2} does not affect these top three rows, we see that u is $-(t')^{rev}$. Now $\Gamma_u^{rev} = \Delta_{t'}$ which concludes the proof.   □

Having identified the Γ_t and $\Delta_{t'}$ that appear in Statement B, let us paraphrase Statement B as follows. If v is a short vector, we may define decorations on $\Gamma_{t(v)}$ and $\Delta_{t(v)'}$. These may be described alternatively as in Chapter 6 or geometrically as in this chapter: b_i is circled if $i = r - 1$ or $2r - 1$ and $b_i = 0$ or if $i \neq r - 1$ or $2r - 1$ and $b_i = b_{i+1}$. Also b_i is boxed if the segment of length b_i that occurs in the canonical path contains the entire i-segment.

Statement B′. *We have*

$$\sum_{\substack{\text{short vector } v \\ \text{wt}(v)=\mu}} G_{w\Gamma}(v) = \sum_{\substack{\text{short vector } v \\ \text{wt}(v)=\mu}} G_{w\Delta}(v),$$

where the sum is over short vectors of a given weight.

This statement is equivalent to Statement B and is thus proved in the preceding chapters.

THEOREM 18.2 *Statement B′ implies Statement A′.*

Proof. This is proved by induction on r. It will perhaps be clearer if we explain this point with a fixed r, say $r = 4$; the general case follows by identical methods. Let b_i and z_i be the entries in $\Gamma(\mathfrak{T}_v)$ and $\Delta(q_r\mathfrak{T}_v)$ as in Proposition 2.2. Thus we have two paths from v to v_{low}:

$$v\begin{bmatrix} b_1 & b_2 & b_3 & b_4 & b_5 & b_6 & b_7 & b_8 & b_9 & b_{10} \\ 1 & 2 & 1 & 3 & 2 & 1 & 4 & 3 & 2 & 1 \end{bmatrix}v_{low} \qquad (18.4)$$

and

$$v\begin{bmatrix} z_1 & z_2 & z_3 & z_4 & z_5 & z_6 & z_7 & z_8 & z_9 & z_{10} \\ 4 & 3 & 4 & 2 & 3 & 4 & 1 & 2 & 3 & 4 \end{bmatrix}v_{low}. \qquad (18.5)$$

We split the first path into two:

$$v\begin{bmatrix} b_1 & b_2 & b_3 & b_4 & b_5 & b_6 \\ 1 & 2 & 1 & 3 & 2 & 1 \end{bmatrix}v', \qquad v'\begin{bmatrix} b_7 & b_8 & b_9 & b_{10} \\ 4 & 3 & 2 & 1 \end{bmatrix}v_{low}.$$

Since $1, 2, 1, 3, 2, 1$ is a reduced decomposition of the long element in the Weyl group of Cartan type $A_3 = A_{r-1}$ generated by the $1, 2, 3$ root operators, v' is the lowest weight vector in the connected component containing v of the subcrystal obtained by discarding the edges labeled r. This means that we may replace

$$v\begin{bmatrix} b_1 & b_2 & b_3 & b_4 & b_5 & b_6 \\ 1 & 2 & 1 & 3 & 2 & 1 \end{bmatrix}v' \qquad \text{by} \qquad v\begin{bmatrix} c_1 & c_2 & c_3 & c_4 & c_5 & c_6 \\ 3 & 2 & 3 & 1 & 2 & 3 \end{bmatrix}v'$$

and we obtain a new path:

$$v\begin{bmatrix} c_1 & c_2 & c_3 & c_4 & c_5 & c_6 & b_7 & b_8 & b_9 & b_{10} \\ 3 & 2 & 3 & 1 & 2 & 3 & 4 & 3 & 2 & 1 \end{bmatrix}v_{low}.$$

We split this again:

$$v \begin{bmatrix} c_1 & c_2 & c_3 \\ 3 & 2 & 3 \end{bmatrix} v'', \qquad v'' \begin{bmatrix} c_4 & c_5 & c_6 & b_7 & b_8 & b_9 & b_{10} \\ 1 & 2 & 3 & 4 & 3 & 2 & 1 \end{bmatrix} v_{\text{low}}.$$

Now $3, 2, 3$ is a reduced word for the long element in the Weyl group of Cartan type $A_2 = A_{r-2}$ generated by the simple reflections except the first and last, σ_1 and $\sigma_r = \sigma_4$. Discarding the edges labeled 1 and 4 from the crystal, it breaks up into A_{r-2} crystals, and v'' may be characterized as the unique short vector in the same connected such A_{r-2} crystal that contains v.

Similarly, starting from (18.5) one may replace the initial $4, 3, 4, 2, 3, 4$ segment by $2, 3, 2, 4, 3, 2$, obtaining

$$v \begin{bmatrix} d_1 & d_2 & d_3 & d_4 & d_5 & d_6 & z_7 & z_8 & z_9 & z_{10} \\ 2 & 3 & 2 & 4 & 3 & 2 & 1 & 2 & 3 & 4 \end{bmatrix} v_{\text{low}}.$$

Due to the characterization of v'' as the unique short vector in the same A_{r-2} crystal as v, this path also goes through v'', by the same argument as with the word $3, 2, 3, 1, 2, 3, 4, 3, 2, 1$. Therefore we have another path from v'' to v_{low}, namely

$$v'' \begin{bmatrix} d_4 & d_5 & d_6 & z_7 & z_8 & z_9 & z_{10} \\ 4 & 3 & 2 & 1 & 2 & 3 & 4 \end{bmatrix} v_{\text{low}}.$$

Now we split the path again:

$$v \begin{bmatrix} d_1 & d_2 & d_3 & d_4 & d_5 & d_6 \\ 2 & 3 & 2 & 4 & 3 & 2 \end{bmatrix} v''', \qquad v \begin{bmatrix} z_7 & z_8 & z_9 & z_{10} \\ 1 & 2 & 3 & 4 \end{bmatrix} v_{\text{low}}.$$

We observe that $2, 3, 2, 4, 3, 2$ is a reduced decomposition of the long element in the Weyl group of Cartan type $A_3 = A_{r-1}$ whose crystals are obtained by discarding edges labeled 1, and so v''' is a lowest weight vector of one of these, so we have also a path

$$v \begin{bmatrix} z_1 & z_2 & z_3 & z_4 & z_5 & z_6 \\ 4 & 3 & 4 & 2 & 3 & 4 \end{bmatrix} v''',$$

which we splice in and now we have obtained the path (18.5) by the following sequence of alterations of (18.4):

$$v \begin{bmatrix} b_1 & b_2 & b_3 & b_4 & b_5 & b_6 & b_7 & b_8 & b_9 & b_{10} \\ 1 & 2 & 1 & 3 & 2 & 1 & 4 & 3 & 2 & 1 \end{bmatrix} v_{\text{low}}$$

$$v \begin{bmatrix} c_1 & c_2 & c_3 & c_4 & c_5 & c_6 & b_7 & b_8 & b_9 & b_{10} \\ 3 & 2 & 3 & 1 & 2 & 3 & 4 & 3 & 2 & 1 \end{bmatrix} v_{\text{low}}$$

$$v \begin{bmatrix} c_1 & c_2 & c_3 & d_4 & d_5 & d_6 & z_7 & z_8 & z_9 & z_{10} \\ 3 & 2 & 3 & 4 & 3 & 2 & 1 & 2 & 3 & 4 \end{bmatrix} v_{\text{low}}$$

$$v \begin{bmatrix} d_1 & d_2 & d_3 & d_4 & d_5 & d_6 & z_7 & z_8 & z_9 & z_{10} \\ 2 & 3 & 2 & 4 & 3 & 2 & 1 & 2 & 3 & 4 \end{bmatrix} v_{\text{low}}$$

$$v \begin{bmatrix} z_1 & z_2 & z_3 & z_4 & z_5 & z_6 & z_7 & z_8 & z_9 & z_{10} \\ 4 & 3 & 4 & 2 & 3 & 4 & 1 & 2 & 3 & 4 \end{bmatrix} v_{\text{low}}$$

These correspond to the words Ω_i where $\Omega_1 = \Omega_\Gamma$, $\Omega_5 = \Omega_\Delta$ and

$$\Omega_2 = (3, 2, 3, 1, 2, 3, 4, 3, 2, 1),$$
$$\Omega_3 = (3, 2, 3, 4, 3, 3, 1, 2, 3, 4),$$
$$\Omega_4 = (2, 3, 2, 4, 3, 3, 1, 2, 3, 4).$$

Let us define boxing and circling rules for each of the five words. In every case, the boxing rule will be that we box an entry if the path segment corresponding to it is an entire root string, as illustrated for Ω_Γ in Figure 2.2. The circling rules for $\Omega_1 = \Omega_\Gamma$ and $\Omega_5 = \Omega_\Delta$ are already given, and in the remaining cases we use the rules in Table 18.1.

Circling rule for Ω_2		Circling rule for Ω_3		Circling rule for Ω_4	
c_1	$c_1 = 0$	c_1	$c_1 = 0$	d_1	$d_1 = d_2$
c_2	$c_2 = c_3$	c_2	$c_2 = c_3$	d_2	$d_2 = d_3$
c_3	$c_3 = 0$	c_3	$c_3 = 0$	d_3	$d_3 = 0$
c_4	$c_4 = c_5$	d_4	$d_4 = d_5$	d_4	$d_4 = d_5$
c_5	$c_5 = c_6$	d_5	$d_5 = d_6$	d_5	$d_5 = d_6$
c_6	$c_6 = 0$	d_6	$d_6 = 0$	d_6	$d_6 = 0$
b_7	$b_7 = b_8$	z_7	$z_7 = z_8$	z_7	$z_7 = z_8$
b_8	$b_8 = b_9$	z_8	$z_8 = z_9$	z_8	$z_8 = z_9$
b_9	$b_9 = b_{10}$	z_9	$z_9 = z_{10}$	z_9	$z_9 = z_{10}$
b_{10}	$b_{10} = 0$	z_{10}	$z_{10} = 0$	z_{10}	$z_{10} = 0$

Table 18.1 Circling rules for Ω_2, Ω_3, and Ω_4.

Now

$$\sum_{\mathrm{wt}(v)=\mu} G_{\Omega_1}(v) = \sum_{\mathrm{wt}(v)=\mu} G_{\Omega_2}(v) = \sum_{\mathrm{wt}(v)=\mu} G_{\Omega_3}(v) =$$
$$\sum_{\mathrm{wt}(v)=\mu} G_{\Omega_4}(v) = \sum_{\mathrm{wt}(v)=\mu} G_{\Omega_5}(v) \quad .$$

The first equality is Statement A′ for A_3, which is the inductive hypothesis. The second equality is Statement B′. The third equality is Statement A′ for the A_2 with root operators 2 and 3; again this is the inductive hypothesis. The third equality is Statement A′ for the A_3 with root operators 2,3,4, again the inductive hypothesis. Putting everything together, we obtain Statement A′ for $r = 4$. The case of general r is similar.   □

Finally, we characterize accordions among preaccordions. Recall that $\alpha_1, \cdots, \alpha_r$ are the simple roots.

PROPOSITION 18.3 *Let v be a short vector in $\mathcal{B}_\lambda$. Then the associated preaccordions $\Gamma_{\mathfrak{T}_v}$ and $\Delta_{\mathfrak{T}'_v}$ are accordions if and only if $\mathrm{wt}(v) - w_0(\lambda)$ is a multiple of the longest root $\alpha_1 + \alpha_2 + \ldots + \alpha_r$.*

Thus the phenomenon of resonance can be understood as relating to the "diagonal" short vectors, whose weights have equal components for all roots.

Proof. By Theorem 18.1 we have

$$v \begin{bmatrix} b_1 & b_2 & \cdots & b_{r-1} & b_r & b_{r+1} & \cdots & b_{2r-1} \\ 1 & 2 & \cdots & r-1 & r & r-1 & \cdots & 1 \end{bmatrix} v_{\text{low}}.$$

This means that the path from v to v_{low} involves $b_1 + b_{2r-1}$ applications of f_1, $b_2 + b_{2r-2}$ applications of f_2, and so forth, and b_r applications of f_r. Since $\text{wt}(f_i(x)) = \text{wt}(x) - \alpha_i$, and since $\text{wt}(v_{\text{low}}) = w_0(\lambda)$, this means that

$$\text{wt}(v) - w_0(\lambda)(b_1 + b_{2r-1})\alpha_1 + (b_2 + b_{2r-2})\alpha_2 + \ldots + b_r \alpha_r.$$

Since the roots are linearly independent, this means $\text{wt}(v) - w_0(\lambda)$ is a multiple of $\alpha_1 + \ldots + \alpha_r$ if and only if $b_1 + b_{2r-1} = \ldots = b_{r-1} + b_{r+1} = b_r$. This is precisely the condition for (18.2) to be an accordion. $\qquad\square$

Chapter Nineteen

Statement B and the Yang-Baxter Equation

The p-parts of Weyl group multiple Dirichlet series, with their deformed Weyl denominators, may be expressed as partition functions of exactly solved models in statistical mechanics. The transition to ice-type models represents a subtle shift in emphasis from the crystal basis representation, and suggests the introduction of a new tool, the Yang-Baxter equation. This tool was introduced by Baxter [1] for proving the commutativity of *row transfer matrices* for the six-vertex and similar models. This is significant for us, because Statement B can be formulated in terms of the commutativity of two row transfer matrices. This last reformulation opens the door to a wide array of possible connections to statistical mechanics and quantum groups that we only begin to survey in this chapter.

In Brubaker, Bump, Chinta, Friedberg, and Gunnells [9], Type A p-parts, or equivalently Type A metaplectic Whittaker functions, are represented as partition functions for the six vertex model. That paper also reformulates Statement B as the commutativity of row transfer matrices, a reformulation that is valid for any metaplectic cover of degree $n \geqslant 1$. However, an alternate proof of Statement B using the Yang-Baxter equation has only been given for $n = 1$. After introducing the terms and tools of this combinatorial approach to statistical mechanics, we will present the complete story for the case $n = 1$, referring to [9] for what is known about general metaplectic covers.

We recall the basics of the *six-vertex model*. This is a class of solvable lattice models, sometimes called *ice-type models* since they may be thought of as describing possible arrangements of water molecules in a two-dimensional grid.

We begin by describing the data needed to specify a model of this class, which we refer to as a *system*. The first ingredient is a planar graph. Let G be the set of vertices of this graph. Every vertex $v \in$ G will have four adjacent edges and appear in one of two possible orientations:

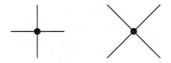

The edges can be classified as *interior edges* (in the middle of the graph) and *boundary edges* (around the perimeter). More precisely, each interior edge joins two vertices, but a boundary edge is adjacent to only one vertex.

The second ingredient needed to specify the model is an assignment of a fixed choice of "spin" $+$ or $-$ to each boundary edge.

The final ingredient needed to specify the model is a map $\mathrm{BW} : \mathrm{G} \to \mathbb{C}^6$. If

$v \in$ G we will write

$$(a_1(v), a_2(v), b_1(v), b_2(v), c_1(v), c_2(v)) = \text{BW}(v).$$

Given these data, we consider all possible assignments of spins $+$ or $-$ to interior edges such that the spins placed on the four edges adjacent to a vertex match those in one of six admissible configurations, listed in Table 19.1. This table assigns to each vertex v one of the six values $a_1(v), a_2(v), b_1(v), b_2(v), c_1(v), c_2(v)$ depending on these adjacent spins. This assigned value is referred to as the *Boltzmann weight* of the vertex v relative to the configuration of spins on the adjacent edges. The spin $\pm$ associated to each edge is written inside a little circle, for easier reading.

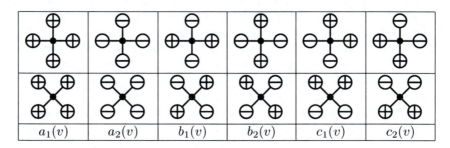

Table 19.1 Admissible configurations at a vertex v, and their Boltzmann weights.

These data – a finite graph, boundary spins on exterior edges, and Boltzmann weights – specify a *system* for the six-vertex model, which we will denote S. Each assignment of spins to every interior edge using only the six admissible configurations of Table 19.1 will be called a *state* $\mathfrak{s}$ of the system. We will write $\mathfrak{s} \in$ S if $\mathfrak{s}$ is a state of the system.

Let $B(v) := B(v, \mathfrak{s})$ denote the assignment of a Boltzmann weight to a vertex labeled v in the state $\mathfrak{s}$ according to Table 19.1. Then we define the Boltzmann weight $B(\mathfrak{s})$ of the state $\mathfrak{s}$ as follows:

$$B(\mathfrak{s}) = \prod_{v \in G} B(v, \mathfrak{s}), \qquad (19.1)$$

where the product is over all vertices of the underlying graph. Finally, we define the *partition function* of the system to be

$$Z(S) = \sum_{\mathfrak{s} \in S} B(\mathfrak{s}).$$

In drawing the graph, we will label each vertex v with $\text{BW}(v)$. Thus two vertices have the same label if they have the same Boltzmann weights. For example, consider the system pictured below consisting of a single row of vertices. Let us assume that every vertex $v \in$ G has the same six-tuple of possible assigned values $\text{BW}(v) = R$. Here $\alpha_i, \beta_i \in \{\pm\}$.

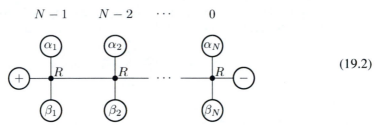

$$(19.2)$$

Let $\alpha = (\alpha_1, \cdots, \alpha_N)$ and $\beta = (\beta_1, \cdots, \beta_N)$ denote the vectors of spins from the top and bottom boundary edges, respectively. For reference we have labeled the columns from 0 to $N - 1$ in ascending order from right to left. Let $\lambda_1 > \lambda_2 > \cdots$ be the numbers of the columns in which $-$ appear in the boundary edges above the row and similarly let $\mu_1 > \mu_2 > \cdots$ be the numbers of the columns in which $-$ appear along the boundary edges below the row. Thus $\lambda = (\lambda_1, \lambda_2, \cdots)$ and $\mu = (\mu_1, \mu_2, \cdots)$ are partitions. In fact they are *strict* partitions, that is, partitions into unequal parts.

For example, if $\alpha = (-, +, +, -, +, -)$ and $\beta = (+, +, -, +, +, -)$ as in the following configuration:

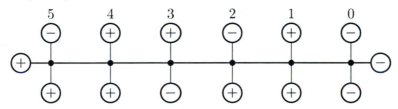

then $\lambda = (5, 2, 0)$ and $\mu = (3, 0)$.

PROPOSITION 19.1 *Given any two vectors of spins α, β, the system (19.2) has either no states or exactly one. Let ℓ and m be the lengths of λ and μ, respectively. Then the system has a state if and only if $m = \ell - 1$ and the partitions λ and μ interleave:*

$$\lambda_1 \geqslant \mu_1 \geqslant \lambda_2 \geqslant \ldots \geqslant \mu_{\ell-1} \geqslant \lambda_\ell.$$

In the example above, the condition is satisfied, and the state is:

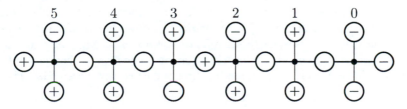

Proof. The number of edges labeled $-$ adjacent to each vertex must be even, so if three edges adjacent to a vertex are determined, the fourth is also determined. From this observation, the interior edges are determined and at most one state can exist.

Let us show that $\lambda_1 \geqslant \mu_1$. The observation that we have just made shows that the spins to the left of the column $\max(\lambda_1, \mu_1)$ occurring in the row are all $+$. If $\mu_1 > \lambda_1$ then the first $-$ among spins below the row is in the μ_1 column, and so in

this column the inadmissible pattern  appears. Therefore $\lambda_1 \geqslant \mu_1$.

Next we claim that $\mu_1 \geqslant \lambda_2$. Suppose that $\lambda_2 > \mu_1$. Then along the row, we have $-$ spins to the right of the λ_1 column up to $\max(\lambda_2, \mu_1)$. Therefore if $\lambda_2 > \mu_1$

the inadmissible pattern  appears in the λ_2 column.

Continuing this way, we see that the partitions λ and μ interleave. It is clear that $m = \ell$ or $\ell - 1$. If $m = \ell$, then the right-most edge spin would have to be $+$, but this is not possible since we have chosen it to be $-$ as part of the fixed boundary conditions of the system.   □

The six-vertex model was shown to be exactly solvable by Lieb [56] and Sutherland [69]. Then Baxter [1] gave a method of evaluating such partition functions based on the notion of *commuting transfer matrices*. To explain this concept, consider a case where the partition function consists of two rows of ice. The set of six Boltzmann weights we assign to each vertex will depend only upon the row in which the vertex occurs. Thus let $R = \mathrm{BW}(v)$ if v is in the first row and $R' = \mathrm{BW}(v)$ if v is in the second row.

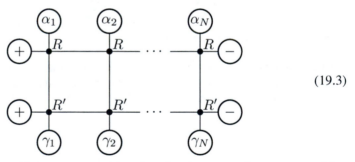

$$(19.3)$$

We may express this two-row partition function in terms of one-row partition functions. Let the partition function of the one-row system in (19.2) be denoted $T(\alpha, \beta)$. We may encode the partition function value for all possible choices of boundary spins α and β in a $2^N \times 2^N$ matrix T_v, where choices of α and β index the rows and columns, respectively. We refer to T as the *row-transfer matrix*. Similarly let T' denote the row transfer matrices for the second row, using the Boltzmann weights R'. Evidently the partition function of the two-row system (19.3) is just

$$\sum_\beta T(\alpha, \beta) T'(\beta, \gamma),$$

where the product is the usual matrix product and the sum ranges over all possible choices of spin vectors β for the vertical edges lying between the two rows.

Note that by Proposition 19.1 the transfer matrix T is nilpotent. This is different from the case with toroidal boundary conditions considered by Baxter [1]. Still we may consider the question of when two transfer matrices commute. That is, for what choices of Boltzmann weights R and R' do we have the identity

$$\sum_\beta T(\alpha, \beta) T'(\beta, \gamma) = \sum_\beta T'(\alpha, \beta) T(\beta, \gamma).$$

This means that the transfer matrix of the two-layer system (19.3) is unchanged if the Boltzmann weights R and R' attached to the vertices of the two rows are interchanged. Baxter gave a sufficient condition for this to occur in terms of what he called the *star-triangle relation*, but which is now referred to as the *Yang-Baxter equation*. We will describe this equation and explain how it can be used to prove the commutativity of two transfer matrices in several cases related to p-parts of multiple Dirichlet series.

DEFINITION 19.2 (Yang-Baxter Equation) *Let R, R', R'' be three sets of Boltzmann weights for the six admissible vertices appearing in Table 19.1. These Boltzmann weights are said to satisfy a Yang-Baxter equation if the following equality of partition functions holds for all possible choices of boundary spins $\varepsilon_1, \cdots, \varepsilon_6 \in \{\pm\}$.*

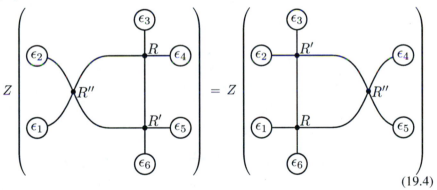

$$(19.4)$$

Note that each admissible state of the system occurring in the partition functions on the left- and right-hand sides of (19.4) is given by a choice of spin on the three internal edges. The equality of these two partition functions is one formulation of the *Yang-Baxter equation*.

PROPOSITION 19.3 *Let T, T' denote transfer matrices for one-row systems of the six-vertex model having left-most boundary spin in the row $+$ and right-most boundary spin in the row $-$ as in (19.2). Let R and R' be their associated sets of Boltzmann weights. Suppose there exists a third set of Boltzmann weights R'' such that the Yang-Baxter equation holds. Let*

$$R'' = (a_1'', a_2'', b_1'', b_2'', c_1'', c_2'').$$

Then $a_1'' TT' = a_2'' T'T$. In particular if $a_1'' = a_2'' \neq 0$ the transfer matrices T and T' commute.

The method of proof is due to Baxter [1].

Proof. Consider the following system:

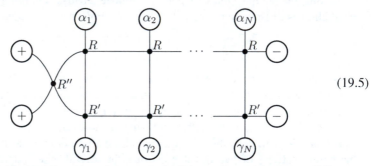

$$(19.5)$$

Consulting Table 19.1 there is only one admissible configuration for the R'' vertex, namely all four edges adjacent to the vertex must be $+$. Therefore the Boltzmann weight for this lone vertex is a_1'' in every admissible state, and this is a constant factor that can be pulled out of the sum defining the partition function. Therefore the partition function of this configuration is a_1'' times the partition function of (19.3). That is, if S is the system (19.3) and S$'$ is the system (19.5) we have

$$Z(\mathsf{S}') = a_1'' Z(\mathsf{S}).$$

On the other hand, using the Yang-Baxter equation, $Z(\mathsf{S}')$ is equal to $Z(\mathsf{S}'')$ where S$''$ is the system:

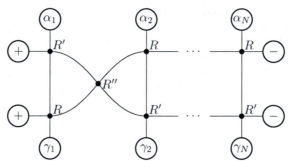

Repeating the process, moving the vertex labeled R'' rightward by means of the Yang-Baxter equation, we eventually see that $a_1'' Z(\mathsf{S}) = Z(\mathsf{S}''')$, where S$'''$ is the configuration:

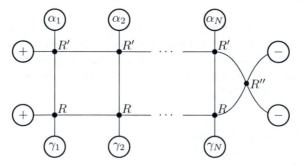

Now the only admissible configuration at the vertex labeled R'' in the above configuration has all four adjacent edges labeled $-$. Therefore the Boltzmann weight at this vertex is a_2'', and we see that the partition function $Z(S''')$ equals a_2'' times the partition function of (19.3) with R and R' reversed. The statement follows. $\quad\square$

We now give examples of particular sets of weights satisfying the Yang-Baxter equation. If $R = (a_1, a_2, b_1, b_2, c_1, c_2)$ we say the Boltzmann weights R are *field-free* if $a_1 = a_2$, $b_1 = b_2$ and $c_1 = c_2$. The field-free case is emphasized in Baxter [1], and was a key example in the development of the theory of quantum groups. If $\mathrm{BW}(v) = R$ we then say that v is *field-free* and write

$$a(v) = a_1(v) = a_2(v), \qquad b(v) = b_1(v) = b_2(v), \qquad c(v) = c_1(v) = c_2(v).$$

Define

$$D(v) = \frac{a(v)^2 + b(v)^2 - c(v)^2}{2a(v)\,b(v)},$$

and if $R = \mathrm{BW}(v)$ we write $D(R) = D(v)$.

THEOREM 19.4 (Baxter) *Suppose that R and R' are field-free sets of Boltzmann weights, and let D be a complex number such that $D(R) = D(R') = D$. Let $R'' = (a'', a'', b'', b'', c'', c'')$ where*

$$a'' = aa' + bb' - 2ab'D,$$
$$b'' = ba' - ab',$$
$$c'' = cc'.$$

Then the Yang-Baxter equation is satisfied.

Proof. This may be checked by direct calculation. $\quad\square$

It may be further shown that R'' also satisfies $D(R'') = D$. Combining Theorem 19.4 and Proposition 19.3, we obtain:

Corollary to Theorem 19.4. *If R and R' are field-free and $D(R) = D(R')$ then the corresponding transfer matrices commute.*

Now let us describe an "ice" version of Tokuyama's formula. We begin by defining a system S_λ^Γ, which depends on a choice of partition $\lambda = (\lambda_1, \cdots, \lambda_{r+1})$. Given λ, we consider a graph having $r + 1$ rows and $\lambda_1 + r$ columns. For reference, we will label the columns beginning with 0 in the right-most column and increasing from right to left. We will label the rows from 1 to $r + 1$, increasing from bottom to top. The spins on the boundary edges will be labeled as follows. On the left and bottom boundary edges, we put $+$. On the right boundary edges, we put $-$. The spins on the top edge are determined by λ. Let $\rho = (r, r - 1, \cdots, 0)$, so $\lambda + \rho = (\lambda_1 + r, \lambda_2 + r - 1, \cdots, \lambda_{r+1})$. We put $-$ on the top boundary edges in the columns numbered with components of $\lambda + \rho$, that is, in columns numbered $\lambda_1 + r, \lambda_2 + r - 1, \cdots, \lambda_{r+1}$. We put $+$ at the remaining top boundary edges.

For example, suppose that $r = 2$ and $\lambda = (3, 1, 0)$. Then $\lambda + \rho = (5, 2, 0)$ and we have the following graph corresponding to the choices of boundary spins described above:

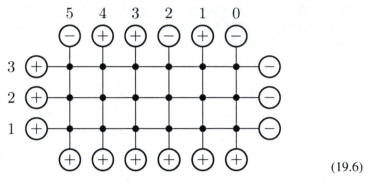

$$(19.6)$$

So in particular, we place a $-$ on the top boundary edges in columns $5, 2$, and 0.

Remark. We note that we have reversed the ordering of the rows from [16]. We have also reversed the order of the variables $z_1, \cdots, z_{r+1}$ in Tokuyama's Theorem 5.1. The reader should bear these changes in mind in comparing the different papers.

To complete the description of the system S_λ^Γ we must specify the Boltzmann weights. We introduce a parameter t that will correspond to the parameter t in Tokuyama's Theorem 5.1. In the i-labeled row of the grid, we will use the following Boltzmann weights:

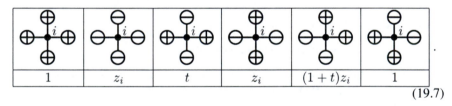

$$(19.7)$$

Remark. Previously we labeled vertices by letters R, R', etc. to denote a corresponding set of Boltzmann weights. Now we are labeling the vertex by an integer i to indicate a set of Boltzmann weights that depend on the index i. It remains true that two vertices with the same label have the same associated sets of Boltzmann weights.

Let $\mathfrak{s}$ be a state of S_λ^Γ. We now explain how to associate to $\mathfrak{s}$ a strict Gelfand-Tsetlin pattern with indexing as in (1.5). From the boundary conditions for S_λ^Γ, there are $r + 1$ columns whose top boundary spin is $-$. Set $a_{00}, \cdots, a_{0r}$ equal to the numbers of these columns in increasing order. From the boundary conditions, we have

$$(a_{00}, \cdots, a_{0r}) = \lambda + \rho.$$

By Proposition 19.1, there are r columns whose spins lying above a vertex in row r are $-$. Set $a_{11}, \cdots, a_{1r}$ equal to the numbers of these columns. Proposition 19.1 asserts that $a_{00} \geqslant a_{11} \geqslant a_{01} \geqslant \cdots \geqslant a_{rr}$. Continuing in this fashion, we obtain a strict Gelfand-Tsetlin pattern, which we will denote $\mathfrak{T}_\mathfrak{s}$, whose entries are the columns where $-$ occur on vertically oriented edges.

For example, given the state of $S^\Gamma_{(3,1,0)}$:

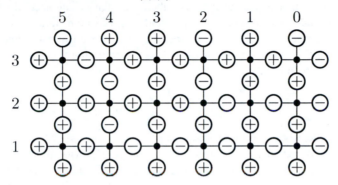

the associated strict Gelfand-Tsetlin pattern is:

$$\left\{ \begin{matrix} 5 && 2 && 0 \\ & 4 && 2 & \\ && 4 && \end{matrix} \right\}.$$

In this example, using the Boltzmann weights as given in (19.7), we have

$$B(\mathfrak{s}) = t^2(1+t)z_1^4 z_2^2 z_3.$$

Note that with the definitions of Chapter 1 we have

$$z^{\mathrm{wt}(\mathfrak{T}_\mathfrak{s})} = z_1^4 z_2^2 z_3, \quad \text{and} \quad G^\flat_\Gamma(\mathfrak{T}_\mathfrak{s}) = t^2(t+1), \quad \text{with } t = -q^{-1}.$$

The product of these two factors is $B(\mathfrak{s})$. The content of the following theorem is that such a relation is always true.

PROPOSITION 19.5 *The map* $\mathfrak{s} \longmapsto \mathfrak{T}_\mathfrak{s}$ *is a bijection between the states* $\mathfrak{s}$ *of* S^Γ_λ *and the set of strict Gelfand-Tsetlin patterns with top row* $\lambda + \rho$. *In the notation of Tokuyama's Theorem 5.1*

$$B(\mathfrak{s}) = (t+1)^{s(\mathfrak{T}_\mathfrak{s})} t^{l(\mathfrak{T}_\mathfrak{s})} z^{\mathrm{wt}(\mathfrak{T}_\mathfrak{s})}.$$

If $t = -q^{-1}$ *then*

$$B(\mathfrak{s}) = G^\flat_\Gamma(\mathfrak{T}_\mathfrak{s}) z^{\mathrm{wt}(\mathfrak{T}_\mathfrak{s})}.$$

Proof. The first statement, the bijection between states and strict Gelfand-Tsetlin patterns, is clear from Proposition 19.1.

We turn to the second statement. First we compare the powers of z_i in $B(\mathfrak{s})$ and in $G^\flat_\Gamma$. We will denote by d_r the sum of the r-th row of $\mathfrak{T}_\mathfrak{s}$.

We note that a vertex in the row labeled i (that is, the i-th row from the bottom) contributes a z_i if there is a $-$ on the edge to the left of the vertex. Therefore the exponent of z_i in $B(\mathfrak{s})$ equals the number of interior edges in that row that have $-$ (not counting the right-most boundary edge, which is $-$ but is not to the left of any vertex).

If $i = 1$, we note that there is a unique vertex in the bottom row that has $-$ above it. It is in the a_{rr} column. The edges in this row to the right of that vertex are therefore $-$. There are $a_{rr} = d_r$ of them, and so the exponent of z_1 is d_r.

Now the $-$ that appear in the row labeled 2, that is, the second row from the bottom, are between columns $a_{r-1,r-1}$ and $a_{r,r}$, or to the right of column $a_{r-1,r}$. Thus the number of $-$ in this row (not counting the right-most boundary edge) is $a_{r-1,r-1} - a_{r,r} + a_{r-1,r}$, that is, $d_{r-1} - d_r$. Continuing this way, we see that the exponent of z_i in $\beta(\mathfrak{s})$ is

$$\begin{cases} d_{r+1-i} - d_{r+2-i} & \text{if } i > 1, \\ d_r & \text{if } i = 1. \end{cases}$$

By (2.17) the product of these factors is $z^{\text{wt}(\mathfrak{T}_\mathfrak{s})}$.

Now let us consider the powers of t. There is a contribution from the vertex v only if there is a $-$ on the edge below v. If the vertex is in the row labeled i and the column labeled $j = a_{ik}$, then it is easy to see that the configuration is:

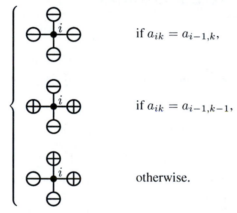

$$\begin{cases} & \text{if } a_{ik} = a_{i-1,k}, \\ \\ & \text{if } a_{ik} = a_{i-1,k-1}, \\ \\ & \text{otherwise.} \end{cases}$$

Therefore the contribution is $1, t$ or $t + 1$, respectively, in the three cases. The product is therefore $(t+1)^{s(\mathfrak{T}_\mathfrak{s})} t^{l(\mathfrak{T}_\mathfrak{s})}$. The last statement of the proposition follows from Lemma 5.2. $\qquad\square$

THEOREM 19.6 *If* $t = -q^{-1}$ *then*

$$Z(S_\lambda^\Gamma) = \sum_{v \in B_{\lambda+\rho}} G_\Gamma^\flat(v) z^{\text{wt}(v)}.$$

Proof. This is an immediate consequence of Proposition 19.5. $\qquad\square$

Corollary to Theorem 19.6. *We have*

$$Z(S_\lambda^\Gamma) = \left[\prod_{i<j} (z_i + t z_j) \right] s_\lambda(z_1, \cdots, z_{r+1}).$$

Proof. This follows from Tokuyama's Theorem 5.1 and Theorem 19.6. $\qquad\square$

Ice versions of Tokuyama's theorem (including a generalization that attaches one deformation parameter t_i to each spectral parameter z_i) were given by Hamel and King [39], [38]. These results were reconsidered in [16], where new proofs were given using the Yang-Baxter equation. In particular, that paper establishes Theorem 19 directly using the Yang-Baxter equation and hence gives a new proof of Tokuyama's theorem.

We have seen that the system of ice S_λ^Γ based on the Boltzmann weights (19.7) corresponds exactly to the $H_\Gamma^\flat$ as in (1.24). We therefore refer to this system as "Gamma ice." Now let us formulate a second system S_λ^Δ, called "Delta ice," that corresponds to $H_\Delta^\flat$ as in (1.25). Given a partition λ, we make use of the same planar graph and boundary conditions for Delta ice as for Gamma ice. The one difference is that for S_λ^Δ, we label the rows in ascending order from top to bottom. The Boltzmann weights are as follows.

⊕ ⊕_j ⊕ ⊕	⊖ ⊖_j ⊖ ⊖	⊖ ⊕_j ⊕ ⊖	⊕ ⊖_j ⊖ ⊕	⊕ ⊖_j ⊖ ⊖	⊖ ⊕_j ⊕ ⊕
1	$z_j t$	1	z_j	$(1+t)z_j$	1

$$(19.8)$$

Again, the label j on the vertex indicates that the set of weights depends on the row numbered j in which it appears. We will indicate the distinction between Gamma ice and Delta ice by denoting the vertices with distinct symbols ● and ○, respectively. Thus for $\lambda = (3, 1, 0)$, the system S_λ^Δ has the following form:

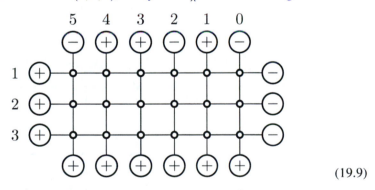

$$(19.9)$$

THEOREM 19.7 *If $t = -q^{-1}$ then*

$$Z(S_\lambda^\Delta) = \sum_{v \in \mathcal{B}_{\lambda+\rho}} G_\Delta^\flat(v) z^{\mathrm{wt}(v)}.$$

Proof. This follows similarly to Theorem 19.6 and we will leave the details to the reader.   □

We may now formulate a version of Statement A in the context of these six-vertex models.

Statement A″. *Given any partition* λ,

$$Z(S_\lambda^\Gamma) = Z(S_\lambda^\Delta).$$

PROPOSITION 19.8 *If* $n = 1$ *then Statement A″ implies Statement A.*

Proof. This is immediate from Theorems 19.6 and 19.7. □

We have stated Theorems 19.6 and 19.7 and Proposition 19.8 for $n = 1$. But in fact all three statements follow for general n via the bijection of Proposition 19.5 using a set of "metaplectic" Boltzmann weights generalizing those given in (19.7) and (19.8). These Boltzmann weights are presented in [9].

In the remainder of this chapter, we will see that the reduction to Statement B also has an exact analog in this context, and we will then prove Statement B (for $n = 1$ only) by means of the Yang-Baxter equation.

In order to reformulate Statement B in the language of our statistical mechanical models, let

$$\boldsymbol{l} = (l_0, \cdots, l_{d+1}) \quad \text{and} \quad \boldsymbol{b} = (b_0, \cdots, b_{d-1})$$

be two strict partitions. Consider the following system having two rows of vertices. In the top row, we use Gamma ice at each vertex with Boltzmann weights as in (19.7). In the bottom row, we use Delta ice with Boltzmann weights as in (19.8). For boundary conditions, we put $+$ on the two left boundary edges and $-$ on the two right boundary edges. We number the columns from right to left as before and put $-$ on the top row in columns whose numbers are in $\boldsymbol{l}$, and $+$ elsewhere; on the bottom row, we put $-$ in the columns whose numbers are in $\boldsymbol{b}$, and $+$ elsewhere. Thus for example if $\boldsymbol{l} = (6, 5, 2, 0)$ and $\boldsymbol{b} = (4, 1)$ then the configuration looks like:

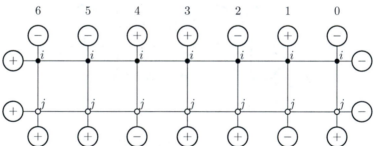

It is a consequence of Proposition 19.1 that states of this system are in bijection with short Gelfand-Tsetlin patterns

$$\mathfrak{t} = \left\{ \begin{matrix} l_0 & & l_1 & & l_2 & & \cdots & & l_{d+1} \\ & a_0 & & a_1 & & \cdots & & a_d & \\ & & b_0 & & \cdots & & b_{d-1} & & \end{matrix} \right\}$$

having fixed top row $\boldsymbol{l}$ and bottom row $\boldsymbol{b}$. Indeed, given such a pattern there is a unique state of the system in which the $-$ in the middle row of vertical edges are in the columns $a_0, a_1, a_2, \cdots$. Thus the short pattern

$$\mathfrak{t} = \left\{ \begin{matrix} 6 & & 5 & & 2 & & 0 \\ & 5 & & 3 & & 0 & \\ & & 4 & & 1 & & \end{matrix} \right\}$$

corresponds to the state

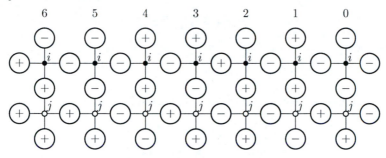

Let us denote this two-row system $S_{l,b}^{\Gamma\Delta}$. Similarly there is a system $S_{l,b}^{\Delta\Gamma}$, which is identical to $S_{l,b}^{\Gamma\Delta}$ except that the Gamma and Delta rows are reversed. For example, with $l = (6, 5, 2, 0)$ and $b = (4, 0)$, we have $S_{l,b}^{\Delta\Gamma}$ as follows.

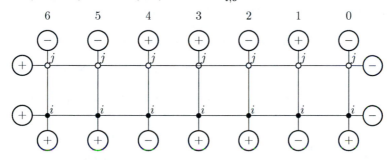

We may now reformulate Statement B in the language of statistical mechanics.

Statement B″. *For strict partitions l and b as above, we have*

$$Z(S_{l,b}^{\Gamma\Delta}) = Z(S_{l,b}^{\Delta\Gamma}).$$

Observe that Statement B″ is precisely the condition that the row transfer matrices for Gamma ice and Delta ice commute. We will prove it below, after discussing the following two results.

PROPOSITION 19.9 *Statement B″ is equivalent to Statement B (when $n = 1$).*

This is extended to the case $n > 1$ in [9].

Proof. We associate with the state $\mathfrak{s}$ of the system $S_{l,b}^{\Gamma\Delta}$ a short Gelfand-Tsetlin pattern $\mathfrak{t}$, in which the entries of the three rows of $\mathfrak{t}$ are the columns with $-$ in the three rows of vertical edges of $\mathfrak{s}$. The top row of $\mathfrak{t}$ is thus l and the bottom row is b, while the middle row depends on $\mathfrak{s}$. Proceeding exactly as in Proposition 19.5 we have

$$B(\mathfrak{s}) = G_\Gamma^b(\mathfrak{t})z^{\mathrm{wt}(\mathfrak{t})},$$

where

$$z^{\mathrm{wt}(\mathfrak{t})} = z_i^{\sum_k (a_k - l_{k+1})} z_j^{\sum_k (a_k - b_k)}.$$

Similarly every state $\mathfrak{s}'$ of $\mathsf{S}^{\Delta\Gamma}_{l,b}$ is associated with such a pattern t' as in (6.4) with

$$B(\mathfrak{s}') = G^b_\Delta(t')z^{\mathrm{wt}(t')},$$

where now

$$z^{\mathrm{wt}(t')} = z_j^{\sum_k (l_k - a'_k)} z_i^{\sum_k (a'_k - b_k)}.$$

Now Statement B″ means that for each monomial $z_i^a z_j^b$ we have

$$\sum_{\mathrm{wt}(t)=z_i^a z_j^b} G^b_\Gamma(t) = \sum_{\mathrm{wt}(t')=z_i^a z_j^b} G^b_\Delta(t').$$

It is easy to check that collecting the terms of equal weight is the same as summing over a prototype, and so this is equivalent to Statement B. $\qquad\square$

THEOREM 19.10 *Statement B″ implies Statement A″.*

A version of the following argument may be found in [16].

Proof. Let us consider the system $\mathsf{S}^\Gamma_\lambda$, illustrated in (19.6). We observe that in the bottom row of vertices, only patterns with a $+$ below can occur, due to the boundary conditions. Referring to the tables in (19.7) and (19.8), we observe that Gamma ice and Delta ice have identical Boltzmann weights for such configurations:

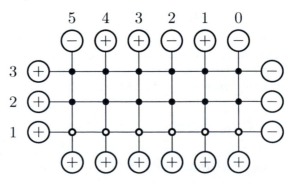

Therefore we may use Delta ice with vertex labels "1" instead of Gamma ice on the bottom row. The system now looks as follows:

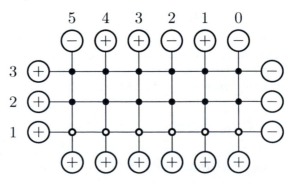

Now we may use Statement B'' to interchange the rows labeled 1 and 2. In fact, we may use it repeatedly, until the 1 row is at the top:

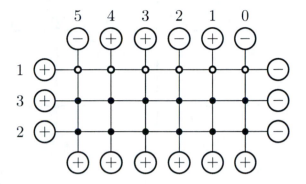

We then repeat the process, changing the row labeled 2 to Delta ice, then using Statement B'' to move it up. Eventually we end up with the configuration (19.9), proving Statement A''. $\square$

Comparing the proof of Theorem 19.10 with the analogous results in Chapter 7, and given as Theorem 18.2, we note that the structures of the proofs are rather similar. Each proof reorganizes the sum, over Gelfand-Tsetlin patterns in Chapter 7, over $\mathcal{B}_{\lambda+\rho}$ in Chapter 18, and over the states of Gamma ice in the present chapter, by repeated applications of the tool at hand. In the Gelfand-Tsetlin or crystal formulations, the tool is the Schützenberger involution, factored into elementary operations that which preserve the sum by repeated applications of Statement B. In the statistical mechanical formulation, the tool is the Yang-Baxter equation in the form of Statement B'', which after repeated applications demonstrates that Gamma ice and Delta ice have the same partition function.

PROPOSITION 19.11 *If $n = 1$, there exists a Yang-Baxter equation proving Statement B''.*

Proof. It is proved in [16] that we have a Yang-Baxter equation as formulated in Definition 19.2 using Boltzmann weights as follows. Set R as in Gamma ice, R' as in Delta ice and let R'' be given according to the following table:

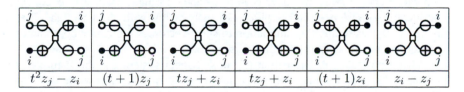

$t^2 z_j - z_i$	$(t+1)z_j$	$tz_j + z_i$	$tz_j + z_i$	$(t+1)z_i$	$z_i - z_j$

We have labeled the edges $\bullet$ and $\circ$ to indicate how they are to be attached to the Gamma ice ($\bullet$) and Delta ice ($\circ$). Thus we claim the following equality of partition functions:

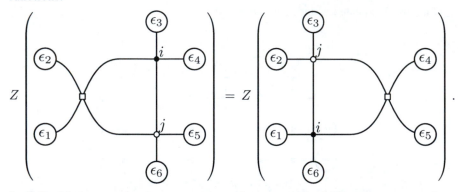

In [16], this is proved as a consequence of a parametrized Yang-Baxter equation with nonabelian parameter group for Boltzmann weights in the free-Fermionic regime. This may also be proved by direct case-by-case comparison for each choice of boundary spins. Given this fact, the proof follows that of Proposition 19.3. □

Though the proofs of various versions of Statement B implying Statement A have the same structure, there is an important difference between the tools used, namely the Schützenberger involution in earlier versions and the Yang-Baxter equation in the present chapter. Proposition 19.5 gives a bijection of the states of the system S_λ^Γ and the set of *strict* Gelfand-Tsetlin patterns with top row $\lambda + \rho$; the set of *all* Gelfand-Tsetlin patterns with top row $\lambda + \rho$ are in bijection with vertices of the crystal $\mathcal{B}_{\lambda+\rho}$. Hence there exists an injection $i : S_\lambda^\Gamma \longrightarrow \mathcal{B}_{\lambda+\rho}$. The image consists of exactly those $v \in \mathcal{B}_{\lambda+\rho}$ for which $G_\Gamma^\flat(v) \neq 0$.

Now let us consider the Schützenberger involution of $\mathcal{B}_{\lambda+\rho}$. This involution is a bijection Sch : $\mathcal{B}_{\lambda+\rho} \longrightarrow \mathcal{B}_{\lambda+\rho}$ with the property that for *most* $v \in \mathcal{B}_{\lambda+\rho}$ we have

$$G_\Gamma^\flat(v) = G_\Delta^\flat(\mathrm{Sch}(v)). \tag{19.10}$$

However this observation does not lead to an easy proof of Statement A. Indeed (19.10) is only true most of the time; to prove that

$$\sum_{\substack{v \in \mathcal{B}_{\lambda+\rho} \\ \mathrm{wt}(v)=\mu}} G_\Gamma^\flat(v) = \sum_{\substack{v \in \mathcal{B}_{\lambda+\rho} \\ \mathrm{wt}(v)=\mu}} G_\Delta^\flat(\mathrm{Sch}(v)), \tag{19.11}$$

one must make sense of the exceptional cases, as we have done.

Statement B is a microcosm for these difficulties. Rather than working with the whole crystal $\mathcal{B}_{\lambda+\rho}$, we have seen in Chapter 18 that certain $v \in \mathcal{B}_{\lambda+\rho}$ could be identified as *short vectors*, which are in bijection with short Gelfand-Tsetlin patterns. By restricting the summation to short vectors, we obtain a combinatorial reduction, but one that still requires further analysis.

When one makes the transition to the statistical system S_λ^Γ, the nature of the problem changes. The bijection Sch is no longer available, since while $i : S_\lambda^\Gamma \longrightarrow \mathcal{B}_{\lambda+\rho}$ is an injection of S_λ^Γ onto most of the crystal $\mathcal{B}_{\lambda+\rho}$, it is not a bijection and its image is not preserved under the Schützenberger involution. Thus although the Schützenberger involution is no longer available as a tool, it is replaced by another powerful tool, the Yang-Baxter equation, and this is sufficient to prove Statement B when $n = 1$.

Although in this chapter we have assumed that $n = 1$, it is shown in [9] that Statement B may be reformulated as the commutativity of row transfer matrices even when $n > 1$. It remains an important open question whether this reformulation may also be proved using some variant of the Yang-Baxter equation. Proposition 13.8 reminds us of the Yang-Baxter equation, for both are cubic relations involving sums of products of three quantities. In [15], it is emphasized that the computation of packets (in the sense of Chapter 13) involves identities of increasing complexity as the length d of the accordion increases. However, it seems these relations can always be reduced to the properties in this proposition, which correspond to Relations (i), (ii), and (iii) in [15]. The fact that this reduction is possible may be evidence of a local relation like the Yang-Baxter equation.

We end this chapter with a remark about resonance, as defined in Chapter 6. We recall that a short Gelfand-Tsetlin prototype is called *totally resonant* if the top row repeats the bottom row, or *resonant* if there are some repetitions between elements of the top row and corresponding elements of the bottom row. Here we observe that resonance has an interpretation in terms of the boundary conditions for $S_{t,b}^{\Gamma\Delta}$ and for $S_{t,b}^{\Delta\Gamma}$. It means that $-$ spins on the top and bottom boundary edges "line up." As a consequence of Theorem 10.1 and the proof of Proposition 19.9, if there is no resonance, the equality of the partition functions follows from a bijection between the admissible states of the two-layer systems, where corresponding states have equal Boltzmann weights. But if there is resonance, we have seen that such a bijection is not possible, and the partition functions must be regarded as subtle superpositions of many states, which miraculously are equal.

Chapter Twenty

Crystals and p-adic Integration

In many cases, integrations of representation theoretic import over the maximal unipotent subgroup of a p-adic group can be replaced by a sum over Kashiwara's crystal $\mathcal{B}(\infty)$. Partly motivated by the crystal description presented in Chapter 2 of this book, this perspective was advocated by Bump and Nakasuji in [21]. Later work by McNamara [65] and Kim and Lee [50] extended this philosophy yet further. Indeed, McNamara shows that the computation of the metaplectic Whittaker function is initially given as a sum over $\mathcal{B}(\infty)$, which, upon closer examination, degenerates to the sum over $\mathcal{B}_{\lambda+\rho}$ presented in Chapter 2. In this chapter, we describe the properties of the crystal $\mathcal{B}(\infty)$, its relation via natural morphisms to the crystals $\mathcal{B}_{\lambda+\rho}$ used earlier, and its role in unipotent integrations related to Whittaker functions.

Let $\widehat{G}$ be a reductive algebraic group over $\mathbb{C}$ with maximal torus $\hat{T}$. The weight lattice Λ is the group of rational characters of $\hat{T}$. We partition the root system Φ into positive and negative roots in the usual way, and let $\{\alpha_1, \cdots, \alpha_r\}$ be the simple roots. A weight λ is *dominant* if $\langle \lambda, \alpha_i \rangle \geqslant 0$ for all α_i. We will denote by Ξ the set of linear combinations of the α_i with nonnegative integer coefficients.

If λ is a dominant weight, let V_λ be the finite-dimensional representation of $\widehat{G}(\mathbb{C})$ with highest weight λ. The multiplicity formula of Kostant [52] expresses the character of V_λ as an alternating sum of translates of a particularly simple generating function by weights $w(\lambda + \rho) - \rho$. This simple generating function may be regarded as the character of the representation of $\hat{T}(\mathbb{C})$ on the enveloping algebra $U(\mathfrak{u}_-)$, where $\mathfrak{u}_-$ is the Lie algebra of the negative maximal unipotent subgroup. This generating function and its translates may alternatively be interpreted as characters of universal modules of $\widehat{G}(\mathbb{C})$ with given highest weight. Thus Verma [71] and Bernstein, Gelfand and Gelfand [6] interpreted the Kostant multiplicity formula as expressing V_λ as an alternating sum of such universal modules.

Taken over to the quantum group, Kostant's generating function is the character of the quantized enveloping algebra $U_q(\mathfrak{u}_-)$. It is associated with a crystal, called $\mathcal{B}(\infty)$ by Kashiwara. See Kashiwara [46], [47], Cliff [32], and Hong and Lee [43].

If λ is dominant, then the character χ_λ of V_λ is an element of $\mathbb{C}[\Lambda]$. We may write

$$\chi_\lambda(z) = \sum_\mu m(\mu, \lambda) z^\mu, \qquad z \in \hat{T}(\mathbb{C}),$$

where $m(\mu, \lambda)$ is the weight multiplicity of μ in λ. We write z^μ for the application of the rational character μ of $\hat{T}$ to $z \in \hat{T}(\mathbb{C})$. The support of the function m is the set of μ in the convex hull of the W-orbit of λ such that $\lambda - \mu$ is in the weight lattice. In particular, the support is contained in $\lambda - \Xi$.

If ξ is an element of the root lattice, let $\mathcal{P}(\xi)$ be the number of *vector partitions* *of* ξ, that is, the number of ways of writing ξ as a nonnegative linear combination of positive roots:

$$\xi = \sum_{\alpha \in \Phi^+} k_\alpha \alpha, \qquad k_\alpha \in \mathbb{Z}_{\geq 0}. \tag{20.1}$$

The function $\mathcal{P}$ is the *Kostant partition function*. It is supported on Ξ. The linear map $\alpha \longmapsto -w_0 \alpha$ permutes the positive roots, from which we deduce that

$$\mathcal{P}(-w_0 \xi) = \mathcal{P}(\xi). \tag{20.2}$$

THEOREM 20.1 (Kostant [52]) *We have*

$$m(\mu, \lambda) = \sum_{w \in W} (-1)^{l(w)} \mathcal{P}(w\lambda + w\rho - \rho - \mu). \tag{20.3}$$

Proof. Let us begin with the Weyl character formula

$$\chi_\lambda(z) = z^{-\rho} \prod_{\alpha \in \Phi^+} (1 - z^{-\alpha})^{-1} \sum_{w \in W} (-1)^{l(w)} z^{w(\lambda + \rho)}.$$

Formally expanding the geometric series, we may rewrite the right-hand side as follows:

$$\sum_{w \in W} (-1)^{l(w)} z^{w(\lambda + \rho) - \rho} \prod_{\alpha \in \Phi^+} \sum_{k_\alpha = 0}^{\infty} z^{-k_\alpha \alpha}.$$

The coefficient of z^μ in

$$z^{w(\lambda + \rho) - \rho} \prod_{\alpha \in \Phi^+} \sum_{k_\alpha = 0}^{\infty} z^{-k_\alpha \alpha}$$

is the number of ways of writing

$$w(\lambda + \rho) - \rho - \mu = \sum_{\alpha \in \Phi^+} k_\alpha \alpha$$

with k_α nonnegative integers, that is, $\mathcal{P}(w\lambda + w\rho - \rho - \mu)$. Summing over w, we obtain (20.3). $\qquad\square$

Now let $T_{-\lambda}$ be the one-dimensional $\hat{T}(\mathbb{C})$-module with character $-\lambda$. Then $V_\lambda \otimes T_{-\lambda}$ is of course not a $\hat{G}(\mathbb{C})$-module, but it is a $\hat{T}(\mathbb{C})$-module, and we may consider its character. This is just the translate of χ_λ by $-\lambda$, that is

$$\text{char}(V_\lambda \otimes T_{-\lambda}) = \sum_\mu m(\mu, \lambda) z^{\mu - \lambda} = \sum_\mu m(\mu + \lambda, \lambda) z^\lambda.$$

If λ runs through a set of weights in the positive Weyl chamber, let us write $\lambda \longrightarrow \infty$ if $\langle \lambda, \alpha_i \rangle \to \infty$ for each simple root α_i. The next result shows that as $\lambda \longrightarrow \infty$, the character $\text{char}(V_\lambda \otimes T_{-\lambda})$ has a limit.

PROPOSITION 20.2 (Kostant [52]) *For any fixed μ, we have*

$$\lim_{\lambda \longrightarrow \infty} m(\mu + \lambda, \lambda) = \mathcal{P}(-\mu).$$

More precisely, given μ,

$$m(\mu + \lambda, \lambda) = \mathcal{P}(-\mu)$$

for all but finitely many dominant weights λ.

Proof. Replacing μ by $\mu + \lambda$ in (20.3) and separating out the term $w = 1$

$$m(\mu + \lambda, \lambda) = \mathcal{P}(-\mu) + \sum_{\substack{w \in W \\ w \neq 1}} (-1)^{l(w)} \mathcal{P}(w\lambda + w\rho - \rho - \mu - \lambda).$$

Observe that $\mathcal{P}(w\lambda + w\rho - \rho - \mu - \lambda) = 0$ unless

$$\lambda - w\lambda \in w\rho - \rho - \mu - \Xi.$$

If $w \neq 1$ this eventually fails as $\lambda \longrightarrow \infty$. $\qquad\square$

Since $\hat{T}(\mathbb{C})$ normalizes $U_-(\mathbb{C})$ it acts on its Lie algebra $\mathfrak{u}_-$ via the adjoint representation, and hence on the universal enveloping algebra $U(\mathfrak{u}_-)$.

PROPOSITION 20.3 *Let $\mu \in \Lambda$. Then μ has nonzero multiplicity in $U(\mathfrak{u}_-)$ if and only if $\mu \in -\Xi$. In this case, the multiplicity of μ in this representation is $\mathcal{P}(-\mu)$.*

Proof. We choose some order on Φ^+. According to the Poincaré-Birkhoff-Witt theorem, a basis for $U(\mathfrak{u}_-)$ consists of the elements

$$\prod_{\alpha \in \Phi^+} X_{-\alpha}^{k_\alpha}$$

where $X_{-\alpha}$ is in the one-dimensional $-\alpha$ eigenspace for the adjoint action of $\hat{T}(\mathbb{C})$ on $\mathfrak{u}_-$, and it is understood that the product is taken in the chosen fixed order on Φ^+. This vector is in the μ weight-space where $-\mu = \sum k_\alpha \alpha$. Hence the dimension of this eigenspace equals the number of vector partitions of $-\mu$. $\qquad\square$

We see from Propositions 20.2 and 20.3 that as the dominant weight $\lambda \longrightarrow \infty$, the module $T_{-\lambda} \otimes V(\lambda)$ of $\hat{T}(\mathbb{C})$ "converges to" $U(\mathfrak{u}_-)$ in the sense that its character converges to the character of $U(\mathfrak{u}_-)$.

We would like to interpret this phenomenon in terms of crystals. The crystal $\mathcal{B}(\infty)$ is a crystal basis of the quantized enveloping algebra of $\mathfrak{u}_-$ (see [59]). We will not pursue this connection here, but rather describe various properties of $\mathcal{B}(\infty)$ and explain how they lead to an analog of the phenomenon that we have just described. For this, we need a more general class of crystals than we used in Chapter 2. Kashiwara gave a definition of crystals that is general enough to include $\mathcal{B}(\infty)$, and our immediate goal is to explain that definition.

For each simple root α_i, let $\alpha_i^\vee$ be the simple coroot. These are linear functionals on the weight lattice Λ, namely $x \longmapsto \frac{2\langle x, \alpha_i \rangle}{\langle \alpha_i, \alpha_i \rangle}$, where $\langle \,, \rangle$ is a W-invariant inner product.

DEFINITION 20.4 *A* **crystal** *of type* Φ *(a rank r root system) is a set C, together with an auxiliary element $0 \notin C$ and the following maps:*

- $e_i, f_i : C \longrightarrow C \cup \{0\}, \quad i = 1, \ldots, r,$

- $\varepsilon_i, \phi_i : C \longrightarrow \mathbb{Z} \cup \{-\infty\}, \quad i = 1, \ldots, r,$

- $\mathrm{wt} : C \longrightarrow \Lambda.$

These maps must satisfy the following conditions:

1. *For any $x, y \in C$, that $e_i(x) = y$ if and only if $f_i(y) = x$. In this case, we further require that*

$$\mathrm{wt}(y) = \mathrm{wt}(x) + \alpha_i, \qquad \varepsilon_i(x) = \varepsilon_i(y) + 1, \qquad \phi_i(x) = \phi_i(y) - 1.$$

2. $\phi_i(x) = \langle \mathrm{wt}(x), \alpha_i^\vee \rangle + \varepsilon_i(x),$ *for all $x \in C$.*

3. *If $\phi_i(x) = -\infty$, then $\varepsilon_i(x) = -\infty$ as well, and $e_i(x) = f_i(x) = 0$.*

If C is a crystal, we associate with C a directed graph with vertices in bijection with elements of C and edges labelled $1, \ldots, r$. We draw an edge $x \xrightarrow{i} y$ whenever $f_i(x) = y$. This is the *crystal graph*.

If $v \in C$ such that $e_i(v) = 0$ for all i, we call v a *highest weight vector*. If the crystal has a unique highest weight vector with weight λ we will usually denote it v_λ.

The functions ϕ_i and ε_i are needed in order to define the tensor product of crystals. We recall that in Chapter 2 we *defined*

$$\phi_i(x) = \max\{k | f_i^k(x) \neq 0\} \quad \text{and} \quad \varepsilon_i(x) = \max\{k | e_i^k(x) \neq 0\} \qquad (20.4)$$

for the particular crystals $\mathcal{B}_\lambda$ that we considered. It is easily checked that they satisfy the required properties above. We say that the crystal C is *normal* if ϕ_i and ε_i are given as in (20.4). However $\mathcal{B}(\infty)$ is not normal. In this case, we will define $\varepsilon_i(x) = \max\{k | e_i^k(x) \neq 0\}$, but the corresponding definition for ϕ_i fails. Instead we will characterize $\mathcal{B}(\infty)$ in terms of tensor products involving highest weight crystals $\mathcal{B}_\lambda$ of Chapter 2.

If C and $\mathcal{D}$ are two crystals, then we define the crystal $C \otimes \mathcal{D}$, the *tensor product* of C and $\mathcal{D}$, as follows. As a set, it is the Cartesian product, but we denote the ordered pair (x, y) with $x \in C$ and $y \in \mathcal{D}$ by $x \otimes y$. We define $\mathrm{wt}(x \otimes y) = \mathrm{wt}(x) + \mathrm{wt}(y)$. Further for $i = 1, \ldots, r$, define

$$f_i(x \otimes y) = \begin{cases} f_i(x) \otimes y & \text{if } \phi_i(x) > \varepsilon_i(y), \\ x \otimes f_i(y) & \text{if } \phi_i(x) \leqslant \varepsilon_i(y), \end{cases}$$

and

$$e_i(x \otimes y) = \begin{cases} e_i(x) \otimes y & \text{if } \phi_i(x) \geqslant \varepsilon_i(y), \\ x \otimes e_i(y) & \text{if } \phi_i(x) < \varepsilon_i(y). \end{cases}$$

It is understood that $x \otimes 0 = 0 \otimes x = 0$. Lastly, we define

$$\phi_i(x \otimes y) = \max(\phi_i(y), \phi_i(x) + \langle \mathrm{wt}(y), \alpha_i^\vee \rangle) \qquad (20.5)$$

and

$$\varepsilon_i(x \otimes y) = \max(\varepsilon_i(x), \varepsilon_i(y) - \langle \mathrm{wt}(x), \alpha_i^\vee \rangle). \qquad (20.6)$$

It may be checked that $\mathcal{C} \otimes \mathcal{D}$ is a crystal with these definitions.

As an example, if λ is any weight, there is a crystal $\mathcal{T}_\lambda$ consisting of a single element t_λ such that $\mathrm{wt}(t_\lambda) = \lambda$. We have $\phi_i(t_\lambda) = \varepsilon_i(t_\lambda) = -\infty$ and $f_i(t_\lambda) = e_i(t_\lambda) = 0$. It is easy to see that if $\mathcal{C}$ is any crystal then for $x \in \mathcal{C}$ we have

$$e_i(x \otimes t_\lambda) = e_i(x) \otimes t_\lambda, \qquad f_i(x \otimes t_\lambda) = f_i(x) \otimes t_\lambda$$

and $\mathrm{wt}(x \otimes t_\lambda) = \mathrm{wt}(x) + \lambda$. So tensoring with $\mathcal{T}_\lambda$ produces a crystal that looks like $\mathcal{C}$, but with the weights shifted by λ.

Given crystals $\mathcal{B}$ and $\mathcal{C}$, a map $\alpha : \mathcal{B} \longrightarrow \mathcal{C} \cup \{0\}$ is called a *morphism* provided the following conditions hold.

1. For any $x, y \in \mathcal{B}$ such that $\alpha(x), \alpha(y) \neq 0$, we have $f_i(x) = y$ if and only if $f_i(\alpha(x)) = \alpha(y)$. (It is equivalent to assume that $e_i(y) = x$ if and only if $e_i(\alpha(y)) = \alpha(x)$.)

2. For all $x \in \mathcal{B}$ with $\alpha(x) \neq 0$, we have $\mathrm{wt}(\alpha(x)) = \mathrm{wt}(x)$, $\phi_i(\alpha(x)) = \phi_i(x)$, and $\varepsilon_i(\alpha(x)) = \varepsilon_i(x)$.

With this definition, crystals become a category. The morphism is called *surjective* if every element of $\mathcal{C}$ is in the image of α.

As above, if λ is a dominant weight we denote by $\mathcal{B}_\lambda$ the normal crystal of tableaux with highest weight λ that was described in Chapter 2.

THEOREM 20.5 (Kashiwara) *Given any root system Φ, there exists a crystal $\mathcal{B}(\infty)$ of type Φ with the following properties:*

1. *$\mathcal{B}(\infty)$ has a unique highest weight vector v_0 and $\mathrm{wt}(v_0) = 0$.*

2. *Given a dominant weight λ, there is a unique surjective morphism $M_\lambda : \mathcal{B}(\infty) \longrightarrow \mathcal{B}_\lambda \otimes \mathcal{T}_{-\lambda}$ such that $M_\lambda(v_0)$ is the highest weight vector $v_\lambda \otimes t_{-\lambda}$ of $\mathcal{B}_\lambda \otimes \mathcal{T}_{-\lambda}$.*

3. *For any $v \in \mathcal{B}(\infty)$, $f_i(v) \neq 0$ and $M_\lambda(f_i(v)) = f_i(M_\lambda(v))$. Moreover if $M_\lambda(v) \neq 0$ then $M_\lambda(e_i(v)) = e_i(M_\lambda(v))$.*

Proof. See Kashiwara [46], Theorem 5 and [47], Theorem 8.1. The map $M_\lambda(v) = \bar{\pi}_\lambda(v) \otimes t_\lambda$, where $\bar{\pi}_\lambda$ is as in [46]; tensoring with t_λ makes v and $M_\lambda(v)$ have the same weight so that M_λ is a morphism. $\qquad \square$

PROPOSITION 20.6 *Let $v \in \mathcal{B}(\infty)$. Then $M_\lambda(v) \neq 0$ for sufficiently large v. Here "sufficiently large" means that the condition becomes true when $\lambda \longrightarrow \infty$.*

Proof. This follows from Lemma 3.2 in Cliff [32], which gives a more precise criterion. $\qquad \square$

Let us imitate the construction of BZL patterns presented in Chapters 2 and 3 for the crystal $\mathcal{B}(\infty)$. Since $\mathcal{B}(\infty)$ has no lowest weight vector, we will use paths along Kashiwara raising operators e_i to the highest weight vector and refer to the resulting path lengths as $\mathrm{BZL}^{(e)}$ patterns, in contrast to our earlier definition on $\mathcal{B}_\lambda$ crystals in (2.8). More precisely, let $\Omega = (\Omega_1, \cdots, \Omega_N)$ be a reduced decomposition of the long element w_0 of W, where $N = l(w_0)$. For $v \in \mathcal{B}(\infty)$ we let $b_1, b_2, b_3, \cdots, b_N$ be nonnegative integers such that $e_{\Omega_j}^{b_j} e_{\Omega_{j-1}}^{b_{j-1}} \cdots e_{\Omega_1}^{b_1} v \neq 0$ but $e_{\Omega_j}^{b_j+1} e_{\Omega_{j-1}}^{b_{j-1}} \cdots e_{\Omega_1}^{b_1} v = 0$. Then we define

$$\mathrm{BZL}_\Omega^{(e)}(v) = (b_1, b_2, \cdots, b_N).$$

If $u = e_{\Omega_N}^{b_N} \cdots e_{\Omega_1}^{b_1} v$ then we write

$$v \left[\begin{array}{ccc} b_1 & \cdots & b_N \\ \Omega_1 & \cdots & \Omega_N \end{array} \right]^{(e)} u.$$

We will prove later that $u = v_0$, the highest weight vector in the crystal $\mathcal{B}(\infty)$.

LEMMA 20.7 *If $v \in \mathcal{B}(\infty)$ and $v' \otimes t_{-\lambda} = M_\lambda(v) \neq 0$, then $e_i(v) \neq 0$ if and only if $e_i(v') \neq 0$, and if this is true then $e_i(v') \otimes t_{-\lambda} = M_\lambda(e_i(v))$.*

Proof. Suppose that $e_i(v) \neq 0$. Let $u = e_i(v)$. Then $f_i(u) = v$, and by Theorem 20.5 we have

$$v' \otimes t_{-\lambda} = M_\lambda(f_i(u)) = f_i M_\lambda(u).$$

Since $v' \otimes t_{-\lambda} \neq 0$ it follows that $M_\lambda(u) \neq 0$ and $e_i(v') \otimes t_{-\lambda} = M_\lambda(u) \neq 0$.

Conversely, suppose that $e_i(v') \otimes t_{-\lambda} \neq 0$. Since M_λ is surjective, there exists $u \in \mathcal{B}(\infty)$ such that $M_\lambda(u) = e_i(v') \otimes t_{-\lambda}$. Since $M_\lambda(u) = e_i(M_\lambda(v))$, and since M_λ is a morphism, $u = e_i(v)$. Therefore $e_i(v) \neq 0$. $\qquad\square$

PROPOSITION 20.8 *Let $v \in \mathcal{B}(\infty)$, and assume that $v' \otimes t_{-\lambda} = M_\lambda(v) \neq 0$. Then $\mathrm{BZL}_\Omega^{(e)}(v) = \mathrm{BZL}_\Omega^{(e)}(v')$. Let $\mathrm{BZL}_\Omega^{(e)}(v) = (b_1, \ldots, b_N)$ and let u be the vertex of $\mathcal{B}(\infty)$ such that*

$$v \left[\begin{array}{ccc} b_1 & \cdots & b_N \\ \Omega_1 & \cdots & \Omega_N \end{array} \right]^{(e)} u.$$

Then $u = v_0$, the highest weight vector in $\mathcal{B}(\infty)$.

Proof. By Lemma 20.7, for all $1 \leqslant j \leqslant N$,

$$e_{\Omega_j}^{b_j} e_{\Omega_{j-1}}^{b_{j-1}} \cdots e_{\Omega_1}^{b_1} v' \neq 0, \qquad e_{\Omega_j}^{b_j+1} e_{\Omega_{j-1}}^{b_{j-1}} \cdots e_{\Omega_1}^{b_1} v' = 0.$$

Therefore $\mathrm{BZL}^{(e)}(v') = (b_1, \cdots, b_N)$. Now by Proposition 2.2 we have $u = v_\lambda \otimes t_{-\lambda}$, the highest weight vector in $V_\lambda \otimes t_{-\lambda}$. Therefore $\mathrm{wt}(u) = 0$. Since $\mathcal{B}(\infty)$ has a unique element with this weight, we have $u = v_0$. $\qquad\square$

Let us illustrate the morphism M_λ of Theorem 20.5 and its properties described above in an example. Below we have drawn the begininning of $\mathcal{B}(\infty)$ when the root system is of Cartan type A_2.

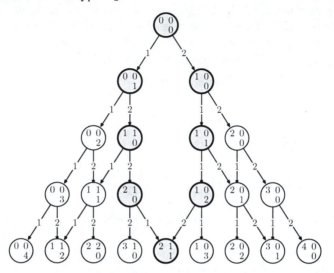

The picture would of course continue downward if we drew more of the infinitely many elements of the crystal. We have used $\Omega = \Omega_\Gamma = (1, 2, 1)$ and, as in the illustrations in Chapter 2 we have labeled each vertex v of the graph by its corresponding BZL pattern $\mathrm{BZL}_\Omega^{(e)}(v)$, laid out in the shape $\begin{pmatrix} b_2 \, b_3 \\ b_1 \end{pmatrix}$.

Let $\lambda = (1, 0, -1)$. Here is the crystal $\mathcal{B}_\lambda$ with highest weight λ:

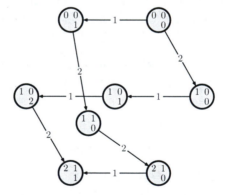

As in Chapters 2 and 3, we have drawn the finite crystal $\mathcal{B}_\lambda$ so that elements of equal weight (e.g., the two vertices in the center of the graph above) are clustered together. We did not draw $\mathcal{B}(\infty)$ this way, opting for the more standard symmetric rendering of the graph. The map $M_\lambda : \mathcal{B}(\infty) \longrightarrow \mathcal{B}_\lambda \otimes \mathcal{T}_{-\lambda}$ can be visualized from these pictures: it sends all unshaded vertices to zero, and it sends the shaded vertex v to the $v' \otimes t_{-\lambda}$, where v' is the vertex of $\mathcal{B}_\lambda$ with the same numbering.

Now let us specialize to the case of Cartan type A_r, and $\Omega = \Omega_\Gamma$ or Ω_Δ in the notation of (2.9), (2.10). In this case, Berenstein and Zelevinsky [5] and Littelmann [57] have computed the set $\{\mathrm{BZL}_\Omega^{(e)}(v) | v \in \mathcal{B}(\infty)\}$.

THEOREM 20.9 *Suppose that* Φ *is of Cartan type* A_r, *and* $\Omega = \Omega_\Gamma$ *or* Ω_Δ. *Let* $\mathfrak{C}$ *be the cone of N-tuples of nonnegative integers* $(b_1, b_2, b_3, \cdots, b_N)$ *such that:*

$$\vdots$$

$$b_4 \geqslant b_5 \geqslant b_6 \geqslant 0$$

$$b_2 \geqslant b_3 \geqslant 0$$

$$b_1 \geqslant 0$$

Then $v \longmapsto \text{BZL}_\Omega^{(e)}(v)$ *is a bijection from* $\mathcal{B}(\infty)$ *to* $\mathfrak{C}$.

Proof. It follows from Propositions 20.6 and 20.8 that the image of $\mathcal{B}(\infty)$ under the map $v \longmapsto \text{BZL}_\Omega^{(e)}(v)$ is the union of the images of $\mathcal{B}_\lambda$ under the maps $\text{BZL}_\Omega^{(e)}$. The statement now follows from Littelmann [57], Theorem 5.1. $\qquad\square$

Littelmann proved more generally that for an arbitrary root system and long word, the set of BZL patterns for $\mathcal{B}(\infty)$ forms a cone.

We now use the explicit description of the cone for $\Omega = \Omega_\Gamma$ or Ω_Δ of Theorem 20.9 to define $G_\Omega^{\flat(e)}$ on $\text{BZL}_\Omega^{(e)}$, the analog of $G_\Omega^\flat$ defined in (2.15). The two definitions are essentially identical, the only difference being that we've now switched to $\text{BZL}_\Omega^{(e)}$ from the patterns $\text{BZL}_\Omega^{(f)}$ of Chapter 2, so the roles of e_i and f_i should be reversed in defining the rules for boxing and circling in patterns, as in Chapter 3.

Note that the boxing rule on $\text{BZL}_\Omega^{(e)}$ is easy: there are no boxes. Indeed, reversing the roles of the raising and lowering operators e_i and f_i in (2.13), the boxing rule for patterns in $\text{BZL}_\Omega^{(e)}$ is as follows. An entry b_k of the pattern is boxed if

$$f_{\Omega_k} e_{\Omega_{k-1}}^{b_{k-1}} \cdots e_{\Omega_1}^{b_1} v = 0.$$

But $f_i(v) \neq 0$ for every $v \in \mathcal{B}(\infty)$, and so this is never satisfied.

The circling rule is also simple. Mimicking the circling rule from Chapters 2 and 3, we circle b_i if it is minimal with respect to cone inequalities with respect to raising operators given in Theorem 20.9. That is, we circle b_i if either i is a triangular number and $b_i = 0$, or if i is not a triangular number and $b_i = b_{i+1}$. Thus the BZL pattern is decorated (with circles only) and we obtain a version of (2.15) applied to the $\mathcal{B}(\infty)$ crystal of the following simple form:

$$G_\Omega^{\flat(e)}(v) = \prod_{b \in \text{BZL}_\Omega^{(e)}(v)} \begin{cases} 1 & \text{if } b_i \text{ is circled,} \\ h^\flat(b_i) & \text{otherwise.} \end{cases} \tag{20.7}$$

For much of this book, we have thought of the generating function built from $G_\Omega^\flat(v)$ on the finite crystal $\mathcal{B}_{\lambda+\rho}$ as the p-part of a multiple Dirichlet series. But as explained in Chapter 4, it may also be viewed as a metaplectic spherical Whittaker function over a p-adic field, evaluated at an element of the maximal torus t_λ. We now explain the analogous result using $G_\Omega^{\flat(e)}(v)$ of (20.7) on $\mathcal{B}(\infty)$.

THEOREM 20.10 *Let us assume that* $|z^\alpha| < 1$ *for positive roots* α. *If* $\Phi = A_r$ *and* $\Omega = \Omega_\Gamma$ *or* Ω_Δ *then*

$$\sum_{v \in \mathcal{B}(\infty)} G_\Omega^{\flat(e)}(v) \, z^{-\text{wt}(v)} = \prod_{\alpha \in \Phi^+} \frac{1 - q^{-1} z^{n\alpha}}{1 - z^{n\alpha}}. \tag{20.8}$$

The right-hand side of (20.8) is precisely the metaplectic version of the Gindikin-Karpelevich formula given in (4.3). It gives the value of the intertwining integral at the identity. We make further comments in this direction after the proof. The theorem was proved by Bump and Nakasuji [21] and McNamara [65]. Another proof was found by Kim and Lee [50]. The three proofs in [21], [65] and [50] are based on different sets of ideas. The argument that we give here is most similar to the proof of Kim and Lee.

Proof. Given any v in $\mathcal{B}(\infty)$, we first note that $G_\Omega^{\flat(e)}(v) \neq 0$ if and only if every entry b_i in $\mathrm{BZL}_\Omega^{(e)}(v)$ is a multiple of n. Indeed, suppose $G_\Omega^{\flat(e)}(v) \neq 0$. Since $h^\flat(k) = 0$ unless $n|k$, this forces all uncircled entries b_i to be divisible by n according to (20.7). If the entry b_j is circled, then either $b_j = b_k$ for some $k > j$ with b_k uncircled (and we apply the previous case) or $b_j = 0$. Hence circled entries must also be divisible by n as well. The converse is immediate.

Let us assume for definiteness that $\Omega = \Omega_\Gamma$; the case $\Omega = \Omega_\Delta$ is the same. We have

$$- \mathrm{wt}(v) = (b_1 + b_3 + b_6 + \ldots)\alpha_1 + (b_2 + b_5 + \ldots)\alpha_2 + \cdots, \qquad \alpha_i\colon \text{simple}.$$

The coefficients of each simple root are the column sums in the array for $\mathrm{BZL}_\Omega^{(e)}(v)$

$$\begin{bmatrix} \ddots & \vdots & \vdots & \vdots & \\ & b_4 & b_5 & b_6 & \\ & & b_2 & b_3 & \\ & & & b_1 & \end{bmatrix}.$$

By Theorem 20.9, the sum on the left-hand side of (20.8) is over integers $b_i, 1 \leq i \leq N = \frac{r(r+1)}{2}$, such that each row is weakly decreasing. Let us produce a vector partition from this data. Specifically, we write

$$- \mathrm{wt}(v) = \sum_{\alpha \in \Phi^+} k_\alpha \alpha$$

where:

$$k_{\alpha_1} = b_1$$
$$k_{\alpha_2} = b_2 - b_3$$
$$k_{\alpha_1 + \alpha_2} = b_3$$
$$k_{\alpha_3} = b_4 - b_5$$
$$k_{\alpha_2 + \alpha_3} = b_5 - b_6$$
$$k_{\alpha_1 + \alpha_2 + \alpha_3} = b_6$$
$$\vdots$$

Observe that for each positive root α, $k_\alpha = b_i$ or $b_i - b_{i+1}$ where the integer i corresponds uniquely to α. Moreover, $k_\alpha = 0$ if and only if b_i is circled. Since $n|b_i$, we have $h^\flat(b_i) = 1 - q^{-1}$. Therefore we have

$$\sum_{v \in \mathcal{B}(\infty)} G_\Omega^\flat(v)\, z^{-\mathrm{wt}(v)} = \prod_{\alpha \in \Phi^+} \sum_{\substack{k_\alpha = 0 \\ n|k_\alpha}}^{\infty} \left\{ \begin{array}{ll} 1 & \text{if } k_\alpha = 0 \\ 1 - q^{-1} & \text{if } k_\alpha > 0 \end{array} \right\} (z^\alpha)^{k_\alpha}.$$

The series is easily evaluated and we obtain (20.8). □

Now let us turn to the formula of Gindikin and Karpelevich and its metaplectic variant that we stated in Chapter 4. Using (4.3) we may write

$$\int_{U(F)} f^\circ(w_0 u)\, du = \sum_{v \in \mathcal{B}(\infty)} G_\Omega^{\flat(e)}(v)\, z^{-\operatorname{wt}(v)}, \tag{20.9}$$

where f° is a spherical vector in the induced representation on the n-fold metaplectic cover of $\operatorname{GL}_{r+1}(F)$, with F a nonarchimedean local field. Here Ω may be either Ω_Γ or Ω_Δ, so $G_\Omega^{\flat(e)}$ may be either $G_\Gamma^{\flat(e)}$ or $G_\Delta^{\flat(e)}$. Observe that $\mathcal{B}(\infty)$ has a symmetry $v \to v'$ such that $G_\Gamma^{\flat(e)}(v) = G_\Delta^{\flat(e)}(v')$ and $\operatorname{wt}(v) = -w_0(\operatorname{wt}(v'))$. So we can equally well write

$$\int_{U(F)} f^\circ(w_0 u)\, du = \sum_{v \in \mathcal{B}(\infty)} G_\Omega^{\flat(e)}(v)\, z^{w_0(\operatorname{wt}(v))}. \tag{20.10}$$

In either (20.9) or (20.10) we have replaced an integration over $U(F)$ by a sum over the crystal $\mathcal{B}(\infty)$.

Recalling the notation of Chapter 4, we may similarly write

$$\int_{U(F)} f^\circ(w_0 u)\psi(t_\lambda u t_{-\lambda})\, du = \sum_{v \in \mathcal{B}_{\lambda+\rho} \otimes \mathcal{T}_{-\lambda-\rho}} G_\Omega^\flat(v) z^{w_0(\operatorname{wt}(v))}. \tag{20.11}$$

For $n = 1$, this follows by combining the Casselman-Shalika formula as in (4.1) with Tokuyama's theorem (Theorem 5.3). Note that the left-hand side of (20.11) is obtained from (4.1) by making a change of variables in u (conjugating by t_λ). The right-hand side of (20.11) correspondingly has weights shifted by $\mathcal{T}_{-\lambda-\rho}$. See Bump and Nakasuji [21] for details. Owing to the surjective morphism $M_{\lambda+\rho} : \mathcal{B}(\infty) \longrightarrow \mathcal{B}_{\lambda+\rho} \otimes \mathcal{T}_{-\lambda-\rho}$, the right-hand side of (20.11) may be pulled back to a sum over $\mathcal{B}(\infty)$. We may even view (20.10) as the result of taking the limit of both sides of (20.11) as $\lambda \to \infty$, though some care needs to be taken in formulating the sense in which this limit is performed; again see [21] for the proof of this fact.

For $n > 1$, the equalities (20.10) and (20.11) remain true when their left-hand sides are interpreted appropriately over the metaplectic group, and are due to McNamara [65]. The paper [65] provides a direct proof of these identities by breaking the unipotent radical into cells over which the spherical function is constant. The cells are determined by an explicit Iwasawa decomposition depending on a choice of long word and are shown to be in bijection with certain subvarieties used by Lusztig [62] to parametrize crystal bases.

Returning to the formula of Gindikin and Karpelevich, the above proof of Theorem 20.10 depends on the fact that the cone inequalities for the BZL patterns $\operatorname{BZL}_\Omega^{(e)}(v)$ suggested a bijection between $\mathcal{B}(\infty)$ and the set of vector partitions. This allows us to interpret the circling rule as follows. The BZL pattern depends on a long word $\Omega = (\Omega_1, \Omega_2, \cdots)$, and this gives a sequential ordering of the positive roots:

$$S(\Omega) = (\alpha_{\Omega_1}, s_{\Omega_1}(\alpha_{\Omega_2}), s_{\Omega_1} s_{\Omega_2}(\alpha_{\Omega_3}), \cdots, s_{\Omega_1} \cdots s_{\Omega_{N-1}}(\alpha_{\Omega_N})).$$

If v corresponds to the vector partition $-\operatorname{wt}(v) = \sum k_\alpha \alpha$ then we circle $b_i \in$ $\operatorname{BZL}_\Omega^{(e)}(v)$ if $k_\alpha = 0$, where α is the i-th positive root in this order.

Therefore, in order to obtain a similar expression for the Gindkin-Karpelevich formula for an arbitrary choice of long word Ω, one needs a suitable rule, depending on Ω, that associates a vector partition with an element of $\mathcal{B}(\infty)$. Such a rule is implicit in Lusztig's canonical basis (see in particular, Lusztig [58, 59, 60, 61] and Kamnitzer [45].) Lusztig defines a graph X whose vertices are the long words, and two vertices are joined by an edge if the corresponding words are related by an application of a braid relation. Then he describes another graph $\tilde{X}$ in which the vertices of X are augmented by data in $\mathbb{N}^\nu$ where ν is the number of positive roots (our N). These are precisely the data needed to define a vector partition, given an ordering of the roots. Now the adjacency relation encodes a transition that describes how to go from a vector partition for the sequence $S(\Omega)$ to a vector partition for the sequence $S(\Omega')$ when the long words Ω and Ω' are related by a braid relation.

If this line of thought is pursued, Lusztig's definition leads to a description of a circling rule and therefore a definition of $G_\Omega^{b(e)}$ that works for $\mathcal{B}(\infty)$ for any long word Ω. By this we mean that

$$\sum_{\mathcal{B}(\infty)} G_\Omega^{b(e)}(v) z^{w_0(v)} = \prod \frac{1 - q^{-1} z^\alpha}{1 - z^\alpha}$$

where, for simplicity, we are limiting ourselves to the case $n = 1$. The works of McNamara [65] and Kim and Lee [50] give some flavor of this.

It would be desirable to give a uniform definition for all long words that similarly works for the Whittaker functions of the kind appearing on the left-hand side of (20.11). An examination of many examples (varying both the type of the root system and their various long element decompositions) shows that such a description will require yet more subtle information beyond the boxing and circling rules of this text. As indicated from our discussion of the Type A case, a more general rule is likely to involve statistics related to the collection of BZL paths and the geometry of their related polyhedra.

Bibliography

[1] R. J. Baxter. *Exactly solved models in statistical mechanics*. Academic Press, Inc., London, 1982.

[2] J. Beineke, B. Brubaker, and S. Frechette. Weyl group multiple Dirichlet series of Type C, preprint, 2010.
http://www-math.mit.edu/ brubaker/beinekebrubakerfrechette.pdf.

[3] J. Beineke, B. Brubaker, and S. Frechette. A crystal definition for symplectic multiple Dirichlet series, preprint, 2010.
http://www-math.mit.edu/ brubaker/edinburgh.pdf.

[4] A. Berenstein and A. Zelevinsky. String bases for quantum groups of type A_r. In *I. M. Gelfand Seminar, Adv. Soviet Math.*, 16:51–89. Amer. Math. Soc., Providence, RI, 1993.

[5] A. Berenstein and A. Zelevinsky. Canonical bases for the quantum group of type A_r and piecewise-linear combinatorics. *Duke Math. J.*, 82(3):473–502, 1996.

[6] I. N. Bernstein, I. M. Gelfand, and S. I. Gelfand. Structure of representations that are generated by vectors of higher weight. *Funckcional. Anal. i Priloen.*, 5(1):1–9, 1971.

[7] S. Bochner. A theorem on analytic continuation of functions in several variables. *Ann. of Math. (2)*, 39(1):14–19, 1938.

[8] B. Brubaker and D. Bump. Residues of Weyl group multiple Dirichlet series associated to $\widetilde{\mathrm{GL}}_{n+1}$. In *Multiple Dirichlet series, automorphic forms, and analytic number theory, Proc. Sympos. Pure Math.*, 75:115–134. Amer. Math. Soc., Providence, RI, 2006.

[9] B. Brubaker, D. Bump, G. Chinta, S. Friedberg, and P. Gunnells. Metaplectic ice, preprint, 2010. http://arxiv.org/abs/1009.1741.

[10] B. Brubaker, D. Bump, G. Chinta, S. Friedberg, and J. Hoffstein. Weyl group multiple Dirichlet series I. In *Multiple Dirichlet series, automorphic forms, and analytic number theory*, Proc. Sympos. Pure Math., 75:91–114, Amer. Math. Soc., Providence, RI, 2006.

[11] B. Brubaker, D. Bump, G. Chinta, and P. Gunnells. Metaplectic Whittaker
      functions and crystals of Type B, preprint, 2010.
      http://sporadic.stanford.edu/bump/spincrystal.pdf.

[12] B. Brubaker, D. Bump, and S. Friedberg. Weyl group multiple Dirichlet se-
      ries. II. The stable case. *Invent. Math.*, 165(2):325–355, 2006.

[13] B. Brubaker, D. Bump, and S. Friedberg. Weyl group multiple Dirichlet se-
      ries, Eisenstein series and crystal bases. To appear in *Ann. of Math.*

[14] B. Brubaker, D. Bump, and S. Friedberg. Weyl group multiple Dirichlet se-
      ries: The stable twisted case. In *Eisenstein series and applications, Progress
      in Math.*, 258:1–26. Birkhäuser, Boston, MA, 2008.

[15] B. Brubaker, D. Bump, and S. Friedberg. Gauss sum combinatorics and meta-
      plectic Eisenstein series. (Dedicated to Steve Gelbart.) In *Automorphic forms
      and L-functions I, Global aspects, Contemp. Math.*, 488:61–81, Amer. Math.
      Soc., Providence, RI, 2009.

[16] B. Brubaker, D. Bump, and S. Friedberg. Schur polynomials and the Yang-
      Baxter equation, preprint, 2010. http://arxiv.org/abs/0912.0911.

[17] B. Brubaker, D. Bump, S. Friedberg, and J. Hoffstein. Weyl group multiple
      Dirichlet series III: Eisenstein series and twisted unstable A_r. *Ann. of Math.*,
      166:293–316, 2007.

[18] A. Bucur and C. Diaconu. Moments of quadratic Dirichlet series over rational
      function fields. *Mosc. Math. Journal*, 10:485–517, 2010.

[19] D. Bump, *Lie groups*. Springer Graduate Texts in Mathematics, vol. 225,
      2004.

[20] D. Bump, S. Friedberg, and J. Hoffstein. On some applications of automor-
      phic forms to number theory. *Bull. Amer. Math. Soc. (N.S.)*, 33(2):157–175,
      1996.

[21] D. Bump and M. Nakasuji. Integration on p-adic groups and crystal bases.
      Proc. Amer. Math. Soc., 138(5):1595–1605, 2010.

[22] W. Casselman. The unramified principal series of p-adic groups. I. The spher-
      ical function. *Compositio Mathematica*, 40:387-406, 1980.

[23] W. Casselman and J. Shalika. The unramified principal series of p-adic groups.
      II. The Whittaker function. *Compositio Math.*, 41(2):207–231, 1980.

[24] G. Chinta. Mean values of biquadratic zeta functions. *Invent. Math.*, 160(1):145–
      163, 2005.

[25] G. Chinta. Multiple Dirichlet series over rational function fields. *Acta Arith.*,
      132(4):377–391, 2008.

[26] G. Chinta, S. Friedberg, and P. Gunnells. On the p-parts of quadratic Weyl group multiple Dirichlet series. *J. Reine Angew. Math.*, 623:1–12, 2008.

[27] G. Chinta, S. Friedberg, and J. Hoffstein. Multiple Dirichlet series and automorphic forms. In *Multiple Dirichlet series, automorphic forms, and analytic number theory, Proc. Sympos. Pure Math.*, 75:3–41. Amer. Math. Soc., Providence, RI, 2006.

[28] G. Chinta and P. Gunnells. Weyl group multiple Dirichlet series constructed from quadratic characters. *Invent. Math.*, 167(2):327–353, 2007.

[29] G. Chinta and P. Gunnells. Constructing Weyl group multiple Dirichlet series. *J. Amer. Math. Soc.*, 23:189–215, 2010.

[30] G. Chinta and P. Gunnells. Littelmann patterns and Weyl group multiple Dirichlet series of type D, preprint, 2009. http://arxiv.org/abs/0909.4558.

[31] G. Chinta and O. Offen. *A metaplectic Casselman-Shalika formula for GL_r*, preprint, 2009. http://www.sci.ccny.cuny.edu/ chinta/publ/mwf.pdf.

[32] G. Cliff. Crystal bases and Young tableaux, *J. Algebra*, 202(1):10–35, 1998.

[33] B. Fisher and S. Friedberg. Sums of twisted GL(2) L-functions over function fields. *Duke Math. J.*, 117(3):543–570, 2003.

[34] B. Fisher and S. Friedberg. Double Dirichlet series over function fields. *Compos. Math.*, 140(3):613–630, 2004.

[35] S. Friedberg. Euler products and twisted Euler products, in *Automorphic forms and the Langlands program. Adv. Lect. Math. (ALM)*, 9:176–198, Int. Press, Somerville, MA, 2010.

[36] S. Friedberg, J. Hoffstein, and Daniel Lieman. Double Dirichlet series and the n-th order twists of Hecke L-series. *Math. Ann.*, 327(2):315–338, 2003.

[37] H. Garland and Y. Zhu. On the Siegel-Weil theorem for loop groups i and ii, preprints, 2008, 2009.
http://arxiv.org/abs/0812.3236 and http://arxiv.org/abs/0906.4749.

[38] A. M. Hamel and R. C. King. U-turn alternating sign matrices, symplectic shifted tableaux and their weighted enumeration. *J. Algebraic Combin.*, 21(4):395–421, 2005.

[39] A. M. Hamel and R. C. King. Bijective proofs of shifted tableau and alternating sign matrix identities. *J. Algebraic Combin.*, 25(4):417–458, 2007.

[40] D. R. Heath-Brown and S. J. Patterson. The distribution of Kummer sums at prime arguments. *J. Reine Angew. Math.*, 310:111–130, 1979.

[41] J. Hoffstein. Eisenstein series and theta functions on the metaplectic group. In *Theta functions: from the classical to the modern, CRM Proc. Lecture Notes*, 1:65–104. Amer. Math. Soc., Providence, RI, 1993.

[42] J. Hong and S.-J. Kang. *Introduction to quantum groups and crystal bases.* *Graduate Studies in Mathematics*, vol. 42, Amer. Math. Soc., Providence, RI, 2002.

[43] J. Hong and H. Lee. Young tableaux and crystal $B(\infty)$ for finite simple Lie algebras. *J. Algebra*, 320(10):3680–3693, 2008.

[44] D. Ivanov, Symplectic Ice. PhD dissertation, Stanford University, 2010.

[45] J. Kamnitzer. Mirković-Vilonen cycles and polytopes.*Ann. of Math.*, 171:245–294, 2010.

[46] M. Kashiwara. On crystal bases of the Q-analogue of universal enveloping algebras. *Duke Math. J.*, 63(2):465–516, 1991.

[47] M. Kashiwara. On crystal bases. In *Representations of groups (Banff, AB, 1994)*, CMS Conf. Proc., 16:155–197, Amer. Math. Soc., Providence, RI, 1995.

[48] M. Kashiwara and T. Nakashima. Crystal graphs for representations of the q-analogue of classical Lie algebras. *J. Algebra*, 165(2):295–345, 1994.

[49] D. A. Kazhdan and S. J. Patterson. Metaplectic forms. *Inst. Hautes Études Sci. Publ. Math.*, (59):35–142, 1984.

[50] H. Kim and K.-H. Lee. Representation theory of p-adic groups and canonical bases, preprint, 2010.
http://www.math.uconn.edu/ khlee/Papers/canonical.pdf.

[51] A. Kirillov and A. Berenstein. Groups generated by involutions, Gelfand-Tsetlin patterns and combinatorics of Young tableaux. *Algebra i Analiz*, 1:92–152, 1995.

[52] B. Kostant. A formula for the multiplicity of a weight. *Trans. Amer. Math. Soc.*, 93:53–73, 1959.

[53] T. Kubota. *On automorphic functions and the reciprocity law in a number field.* Lectures in Mathematics, Department of Mathematics, Kyoto University, No. 2. Kinokuniya Book-Store Co. Ltd., Tokyo, 1969.

[54] R. Langlands. *Euler products.* Yale University Press, New Haven, CT, 1971.
http://publications.ias.edu/rpl/paper/37.

[55] C. Lenart. On the combinatorics of crystal graphs. I. Lusztig's involution. *Adv. Math.*, 211(1):204–243, 2007.

[56] E. Lieb. Exact solution of the problem of entropy in two-dimensional ice. *Phys. Rev. Lett.*, 18:692–694, 1967.

[57] P. Littelmann. Cones, crystals, and patterns. *Transform. Groups*, 3(2):145–179, 1998.

[58] G. Lusztig. *Introduction to quantum groups. Progress in Math.*, 110. Birkhäuser Boston Inc., Boston, MA, 1993.

[59] G. Lusztig. Canonical bases arising from quantized enveloping algebras. *J. Amer. Math. Soc.*, 3(2):447–498, 1990.

[60] G. Lusztig. Canonical bases arising from quantized enveloping algebras. II. In Common trends in mathematics and quantum field theories (Kyoto, 1990). *Progr. Theoret. Phys. Suppl.* 102:175–201, 1990.

[61] G. Lusztig. Introduction to quantized enveloping algebras. In *New developments in Lie theory and their applications* (Cordoba, 1989). *Progress in Math.*, 105:49–65. Birkhäuser Boston, Boston, MA, 1992.

[62] G. Lusztig. An algebraic-geometric parametrization of the canonical basis. *Adv. Math.* 120 (1996), no. 1, 173–190.

[63] I. G. Macdonald. *Spherical functions on a group of p-adic type.* Ramanujan Institute, Centre for Advanced Study in Mathematics, University of Madras, Madras, 1971. Publications of the Ramanujan Institute, No. 2.

[64] I. Macdonald. *Symmetric functions and Hall polynomials*, 2nd ed., Oxford Mathematical Monographs, New York, 1995.

[65] P. J. McNamara. Metaplectic Whittaker functions and crystal bases. *Duke Math. J.*, forthcoming.

[66] M.-P. Schützenberger. La correspondance de Robinson. In *Combinatoire et représentation du groupe symétrique (Actes Table Ronde CNRS, Univ. Louis-Pasteur Strasbourg, Strasbourg, 1976)*, 59–113. Lecture Notes in Math., Vol. 579. Springer, Berlin, 1977.

[67] T. Shintani. On an explicit formula for class-1 "Whittaker functions" on GL_n over P-adic fields. *Proc. Japan Acad.*, 52(4):180–182, 1976.

[68] R. Stanley. *Enumerative combinatorics. Vol. 1, Cambridge Studies in Advanced Mathematics*, vol. 49. Cambridge University Press, Cambridge, 1997.

[69] B. Sutherland. Exact solution for a model for hydrogen-bonded crystals. *Phys. Rev. Lett.*, 19(3):103–104, 1967.

[70] T. Tokuyama. A generating function of strict Gelfand patterns and some formulas on characters of general linear groups. *J. Math. Soc. Japan*, 40(4):671–685, 1988.

[71] D.-N. Verma. Structure of certain induced representations of complex semisimple Lie algebras. *Bull. Amer. Math. Soc.*, 74:160–166, 1968.

Notation

e_i, f_i	Kashiwara operators	10
$\mathrm{Sch} : \mathcal{B}_\lambda \to \mathcal{B}_\lambda$	Schützenberger involution	11
$\psi_\lambda, \phi_\lambda : \mathcal{B}_\lambda \to \mathcal{B}_{-w_0\lambda}$	involutions	11
w_0	long Weyl group element	11
rev	mirror image array	12
ϕ_i, ϵ_i	number of times f_i or e_i applies	13
Ω	a reduced word	14
$v \begin{bmatrix} b_1 & \cdots & b_N \\ \Omega_1 & \cdots & \Omega_N \end{bmatrix} v'$	path from v to v'	14
v_{high}	highest weight vector	14
v_{low}	lowest weight vector	14
$\mathrm{BZL}_\Omega(v), \mathrm{BZL}_\Omega^{(f)}(v)$	string vector of v for long word Ω	14
Ω_Γ	the word $(1, 2, 1, 3, 2, 1, \cdots)$	15
Ω_Δ	the word $(r, r-1, r, r-2, r-1, r, \cdots)$	15
$G_\Gamma(v),\ G_\Delta(v),$		
$\quad G_\Gamma^{(f)}(v),\ G_\Delta^{(f)}(v)$	products of Gauss sums	19
$G_\Gamma^\flat(v), G_\Delta^\flat(v)$	reduced versions of G_Γ, G_Δ	19
$\hat{G}(\mathbb{C})$	$GL_{r+1}(\mathbb{C})$, the L-group	26
$\hat{T}(\mathbb{C})$	diagonal torus in $\hat{G}(\mathbb{C})$	26
G, T, B, U	GL_{r+1}, maximal torus,	
	Borel, maximal unipotent	26
F	nonarchimedean local field	26
z	element of $\hat{T}(\mathbb{C})$	26
χ_z	unramified character	26
W	Weyl group	26
F	nonarchimedean local field	26
δ	modular character of $B(F)$	26
V_χ, π_χ	principal series representation	26
K	$GL_{r+1}(\mathfrak{o})$	26
f°	spherical vector	26
z^μ	value of a weight μ at z	27
ψ_F	character of F	27
ψ	character of $U(F)$	27
W	spherical Whittaker function	28
$\boldsymbol{F}$	a global field	28
$\mathbb{A}$	adele ring of $\boldsymbol{F}$	28
E_ζ	Eisenstein series	29
$\psi_\mathbb{A}$	character of $\mathbb{A}/\boldsymbol{F}$	29
s_λ	Schur polynomial	32
$GT(\lambda), GT_\lambda$	Gelfand-Tsetlin patterns with	
	with top row λ	32
$g^\flat, h^\flat$	$g^\flat(a) = -q^{-1}, h^\flat(a) = (q-1)q^{-1}$	
	$(n = 1 \text{ only})$	34

Index

accordion, 43, 44
   as lattice point, 45
   crystal interpretation, 113
   non-strict, 43
admissibility
   Γ- and Δ-, 97
admissible configuration, 116

belongs to prototype, 38
block
   $\square$- and $\bigcirc$-, 99, 100
Boltzmann weight, 116
bond
   moving, 56
bond-marked cartoon, 55, 57, 64
   indexing and, 65
   resonances and, 59
boundary edge, 115
box-circle duality, 23
boxing, 4, 5, 10, 17, 23, 39
branching rule, 11, 14, 32

canonical basis, 142
canonical indexings, 59, 64, 71
cartoon, 41, 54
   bond-marked, 55, 57, 64
   indexing and bond-marked, 65
   involution and, 55
   preaccordions and, 55
   resonances and bond-marked, 59
   simple, 59, 60, 64
Casselman-Shalika formula, 27, 29, 31, 141
circling, 4, 5, 10, 17, 23, 39
circling compatibility condition, 43
Circling Lemma, 63
Class I, II, III, and IV, 60
coefficient
   reduced, 8, 9, 19, 34
compatible signature, 43, 45, 47

concur, 96
concurrence, 50, 96
   equivalence relation, 96
   operations on signatures and, 97
condition
   circling compatibility, 43
coroots, 134
crystal, 10, 134
   standard, 13
crystal base, 10
crystal graph, 10, 135
crystals
   category of, 136
   morphism, 132, 136
crystals
   tensor product, 13, 135
cut-and-paste simplex, 47

decoration, 4, 17, 18, 39, 43–45, 108, 111
Delta ice, 125
distinguished edge, 56
dominant weight, 10, 132

effective dominant weight, 10
Eisenstein series, 29
episode, 41, 55, 58, 60
   classes of, 60
   indexing and, 58
equalized pair, 100–102

facet, 48, 49, 89, 95
   closed, 48
   open, 48
field-free, 121

Gamma ice, 125
Gauss sum, 4
Gelfand-Tsetlin pattern, 2
   non-strict, 5

Lightning Source UK Ltd.
Milton Keynes UK
UKOW04f1153170414

230149UK00003B/29/P